I0051312

EARNED VALUE MANAGEMENT SYSTEM MANUAL

LGM INTERNATIONAL, LLC

2nd Edition

By Dr. Sean Thomas Regan, FAACE, CCP, CEP, EVP, MRICS, Fellow Project Controls Guild

LGM
International

Development
Construction
Real Estate
Finance

Acknowledgments:

This guide was developed based on the support of numerous companies and people who helped implement the system for numerous companies. I specifically would like to thank Jim Zack, for convincing me to publish. Pete Griesmyer of the AACEI Education Board who stands by my development and beliefs in Project Controls as well as Professor John Hannon of the University Southern Mississippi for the years of support in becoming a professor and convincing me that a practitioner comes from both academia and the field. Alexander Tsvetkov of PMSOFT Group of Companies in Moscow for the educational opportunities.

Then to my Grandfather Robert Regan who was a Project Manager world-wide with Bechtel and Stone and Webster, who left me great insight from the Manhattan Project to Rancho Seco Nuclear Power Plant. My Uncle John Randall (Randy) Regan for the first step into Bechtel and was there for years, to the cognac and cigars in Moscow! My Father Ronald Terry Regan, a California based General Contractor who made me work in the field and appreciate the skills of management of construction. And last and most important to my wife Nataliya, my sons Brendan, Corey, Aidan, and Brian, and Step Daughters Julia and Dariia. You all were with me good and bad through move after move in the construction and project management field. This document would not have come to implementation without your love and support as I took you around and around the world.

The Context of this system manual is a reference guide, it was the authors belief, that by applying a company name versus a generic name, it would allow the users to clearly understand that these processes are a basis for procedures and support of the Earned Value Management System.

**Earned Value Management
System Manual 2nd Edition**
Title ID: 3948870
ISBN-13: 978-0615676210

Copyright ©2012-2017, LGM International, LLC

Use or disclosure of data contained on this sheet is
subject to the restriction on the title page of this document.

Page ii

TABLE OF CONTENTS

1. Introduction
1.1 LGM INTERNATIONAL Organization 2
1.2 Overview 2
1.3 Content..2
1.4 System Objective 3
1.5 System Organization..3
1.6 Compliance 4
1.7 LGM INTERNATIONAL EVMS Internal Surveillance and Surveillance of Subcontractors 4

2. Acronyms and Definitions
2.1 Acronyms and Abbreviations 9
2.2 LGM INTERNATIONAL Earned Value Management System Glossary 10

3. Proposal/Negotiation
3.1 Overview 19
3.2 Establishing the Proposal Plan..19
3.2.1 Initial Planning 19
3.2.2 Critical Proposal Elements 19
3.3 Proposal/Negotiation Process 21
3.3.1 Business Management Decision Process..21
3.3.2 Establishing a Proposal Team 21
3.3.3 Proposal Definition 21
3.3.4 Estimating the Proposal..21
3.3.5 Proposal Review..22

4. Organization
4.1 Overview 25
4.2 Contract Work Breakdown Structure..25
4.2.1 Identification of Contract Work Breakdown Structure 25
4.2.2 CWBS Development 26
4.2.3 Contract Work Breakdown Structure Coding 26
4.2.4 Contract Work Breakdown Structure Index and Dictionary 26
4.3 Organizational Breakdown Structure..26
4.3.1 Organization 26
4.3.2 Organization Coding 27
4.3.3 Project and Functional Organization 27
4.4 Responsibility Assignment Matrix 28
4.4.1 Interrelationship of CWBS and OBS 28
4.4.2 Major Subcontractors 28
4.5 Development of the CWBS Dictionary and Responsibility Assignment Matrix 28
4.6 Management System Integration 29
4.7 Control Account Management 29
4.7.1 Control Account 29
4.7.2 Cost Collection 30
4.7.3 Types of Effort 30
4.7.4 Work Package 30

5. Project Budget Planning
5.1 Overview 38

5.2   Proposal/Budget Relationships 38
5.3   Establishing the Contract Baseline 38
5.4   Budget Baseline Management 39
5.5   Budget Definitions 39
5.6   Control Account Development 40
5.7   Control Account Planning 41
5.8   Management Reserve 41
5.9   Work Package Budgets 41
5.10   Planning Packages 41
5.11   Rolling Wave Planning 41

6.   Scheduling
6.1   Overview 47
6.2   Schedule Hierarchy 47
6.2.1   Baseline Schedules 47
6.2.2   Special Schedules 48
6.3   Scheduling Techniques 48
6.4   Schedule Traceability 48
6.4.1   Vertical Traceability 48
6.4.2   Horizontal Traceability 49
6.4.3   Historical Traceability 49
6.5   Proposal Schedule 49
6.6   Schedule Development 49
6.7   Schedule Statusing and Forecasting and Reviews 50
6.8   Schedule Changes 51
6.9   Schedule Development Process 51
6.10   Project Risk Analysis 51

7.   Work Authorization
7.1   Overview 63
7.2   Work Authorization 63
7.2.1   Contract Work Authorization 63
7.2.2   Project Work Authorization 63
7.2.3   Control Account Work Authorization 63
7.3   Work Authorization Flowchart 64

8.   Performance Measurement
8.1   Overview 75
8.2   Principal Characteristics 75
8.3   Development of Performance Measurement 75
8.4   Performance Measurement Techniques 75
8.4.1   Discrete Effort 76
8.4.2   Apportioned Effort 77
8.4.3   Level of Effort 77
8.5   Control Account Plan 78

9.   Material Planning and Control
9.1   Overview 83
9.2   Objectives 83
9.3   Material Types 83

9.3.1   Material Categories 83

9.4   Material Planning and Control Process...83
9.4.1   Material Estimating 83
9.4.2   Material Budgets 84
9.4.3   Performance Measurement..84
9.4.4   Budget Changes 85
9.4.5   Material Receipt and Usage Analysis 85

9.5   Physical Accountability 85

9.6   Material System Flowchart 86

10.   Subcontract Planning and Control

10.1   Overview 90

10.2   Major Subcontract Definition 90

10.3   Major and Non-Major Subcontracts 90
10.3.1   Flowdown...90
10.3.2   Reviews..90
10.3.3   Surveillance of Subcontractors EVMS...91
10.3.4   Major and Non-Major Subcontracts Reporting 91
10.3.5   Non-Major and Minor Subcontract Reporting 92

10.4   Subcontract Planning, Performance Measurement, and Actual Cost.............................92
10.4.1   Subcontract Planning: Budgeted Cost for Work Scheduled..92
10.4.2   Subcontract Performance Measurement: Budgeted Cost for Work Performed 93
10.4.3   Subcontract Actual Cost of Work Performed 93

10.5   Subcontractor Variance Analysis and Estimates at Completion 94

11.   Cost Accounting

11.1   Overview 99

11.2   Cost Accounting Policy 99

11.3   Cost Accounting Summarization 100

11.4   Accounting Code Structure 100
11.4.1   Organization Codes 101
11.4.2   Accounting Codes 101
11.4.3   Project Codes 101

11.5   Recording Cost 101
11.5.1   Direct Labor 101
11.5.2   Material and Subcontract Costs..102
11.5.3   Other Direct Costs 102
11.5.4   Indirect Costs 102

11.6   Accounting Adjustments...102

12.   Overhead Cost Management

12.1   Overview 106

12.2   Indirect Rates...106
12.2.1   Calculation of Indirect Rates..106
12.2.2   Cost Identification to Pools 107
12.2.3   Allocation to Contract 107
12.2.4   Overhead Application to CWBS 107

12.3   Base Development 107
12.3.1   Preparation of Forecasts 107
12.3.2   Operating Expenses 107
12.3.3   Corporate Accounting Review of Forecasts 107
12.3.4   President's Review of Final Forecasts 107
12.3.5   Approved Budget 108

12.4   Overhead Budget Control 108

13.   Data Analysis and Reporting

13.1   Overview 110
13.2   Data Update 110
13.3   Monthly Reporting Cycle 110
13.3.1 Control Account Manager Responsibilities 111
13.3.2 Project Office Responsibilities 111
13.4   Variance Analysis 111
13.4.1 Variance Analysis Thresholds 111
13.4.2 Variance Types 111
13.4.3 Variance Reporting 112
13.5   Estimate at Completion 112
13.5.1 Comprehensive EAC 113
13.5.2 EAC Updates 113
13.6   Performance Reporting 114
13.6.1 Internal Reporting 114
13.6.2 Customer Reporting 114
13.7   Performance Measurement, Reporting, and Analysis Flowcharts 115
13.7.1 Performance Measurement, Analysis, and EAC Update Flowchart 115
13.7.2 Customer Reporting Flowchart 116
13.7.3 Comprehensive EAC Update Flowchart 117

14.   Revisions and Baseline Control

14.1   Overview 136
14.2   Change Management Process and Documentation 136
14.2.1 Project Change Logs 136
14.3   Contract Changes 137
14.3.1 Negotiated Changes 138
14.3.2 Authorized Un-priced Changes 138
14.4   Replanning Process 139
14.4.1 Internal Replanning 140
14.4.2 Control Account Scope Change 140
14.4.3 Changes to Budget or Schedule to Reflect Efficiency 140
14.4.4 Conversion of Planning Packages into Work Packages 140
14.4.5 Changes Because of Make Versus Buy Decisions 141
14.4.6 Replanning Work Packages 141
14.5   Replanning Requiring Project Manager Approval 141
14.5.1 Major Replanning 141
14.5.2 Formal Reprogramming 141
14.6   Retroactive Adjustments 142

15.   Evaluation/Demonstration Review Checklist for EVMS

15.1   Criteria Checklist Cross Reference 149

Use or disclosure of data contained on this sheet is
subject to the restriction on the title page of this document.

Page vi

LIST OF FIGURES

Figure 1-1   LGM INTERNATIONAL Organizational Chart 6
Figure 1-2   LGM INTERNATIONAL EVMS Overview Flowchart 7
Figure 3-1   Proposal/Negotiation Flowchart 23
Figure 4-1   CWBS and RAM Development Flowchart 31
Figure 4-2   Sample Contract Work Breakdown Structure 32
Figure 4-3   Sample CWBS Index 33
Figure 4-4   Sample CWBS Dictionary 34
Figure 4-5   Sample Project Organization 35
Figure 4-6   Sample Responsibility Assignment Matrix 36
Figure 4-7   Management System Integration 36
Figure 5-1   Budget Distribution Flowdown 43
Figure 5-2   Sample Responsibility Assignment Matrix 44
Figure 5-3   Rolling Wave Planning 45
Figure 6-1   Schedule Hierarchy 54
Figure 6-2   Sample Project Master Schedule 55
Figure 6-3   Sample Intermediate Schedule 56
Figure 6-4   Sample Detailed Schedule 57
Figure 6-5   Schedule Integration and Traceability 58
Figure 6-6   Schedule Development Flowchart 59
Figure 6-8   Project Risk Analysis Process Diagram 61
Figure 7-1   Work Authorization Process 67
Figure 7-2   Sample Contract Work Authorization 68
Figure 7-3   Sample Project Work Authorization 69
Figure 7-4   Sample Control Account Work Authorization 70
Figure 7-5   Sample Control Account Plan 71
Figure 7-6   Work Authorization Flowchart 72
Figure 7-7   Sample CAWA Log 73
Figure 8-1   Sample Work Breakdown Structure Hierarchical Relationships 79
Figure 8-2   Performance Measurement Methods 80
Figure 8-3   Sample Control Account Plan 81
Figure 9-1   Material Subsystem Flowchart 88
Figure 10-1   Minor, Non-Major, and Major/Critical Subcontract Definition 95
Figure 10-2   Subcontract Reporting Requirements 96
Figure 10-3   Technical and/or Schedule Risk Factors 97
Figure 11-1   Accounting Process 103
Figure 11-2   Corporate Coding Structure 104
Figure 13-1   Monthly Reporting Cycle 121
Figure 13-2   Sample Monthly Performance Report 122
Figure 13-3   Sample Variance Analysis Report 123
Figure 13-4   Sample Corrective Action Log 124
Figure 13-5   Sample Control Account ETC PLANNING Sheet 125
Figure 13-6   Monthly Performance Data 126
Figure 13-7   Cost Performance Report, Format 1—Work Breakdown Structure 127
Figure 13-8   Cost Performance Report, Format 2—Organizational Category 128

Use or disclosure of data contained on this sheet is
subject to the restriction on the title page of this document.

Page vii

Figure 13-9   Cost Performance Report, Format 3—Baseline 129

Figure 13-10   Cost Performance Report, Format 4—Staffing 130

Figure 13-11   Cost Performance Report, Format 5—Explanation and Problem Analysis.................... 131

Figure 13-12   Performance Measurement, Analysis, and EAC Update Flowchart............................... 132

Figure 13-13   Customer Reporting Flowchart 133

Figure 13-14   Comprehensive EAC Flowchart.. 134

Figure 14-1   Sample Contract Budget Base Log 143

Figure 14-2   Sample Undistributed Budget Log.. 144

Figure 14-3   Sample Management Reserve Log... 145

Figure 14-4   Contract Change Flowchart 146

Figure 14-5   Internally Generated Change Flowchart 147

Use or disclosure of data contained on this sheet is
subject to the restriction on the title page of this document.

Page viii

1. Introduction

Use or disclosure of data contained on this sheet is
subject to the restriction on the title page of this document.

Page 1

1.1 LGM INTERNATIONAL Organization

LGM INTERNATIONAL is a wholly owned subsidiary of LGM INTERNATIONAL, LLC consists of the Office of the President (President and CEO, Executive Vice President–Operations and Executive Vice President–Finance and Administration), corporate staff, and operational units. Figure 1-1 provides an overview of the LGM INTERNATIONAL organization Operational units are product lines (profit centers). LGM INTERNATIONAL product lines are aligned with the type of industry with which they typically contract. Each product line's organization usually includes the Vice President, product line staff, and the operating departments. All work is contracted through product lines.

The Product Line Vice President identifies a Project Manager for a major contract. If the customer requires the application of the Earned Value Management System (EVMS) in accordance with U.S. Department of Defense (DoD) DoDI 5000.2R, National Aeronautics and Space Administration NPD 9501.3, Department of Energy M 413.3-1, or the Federal Aviation Administration T 1.13, LGM INTERNATIONAL uses the LGM INTERNATIONAL EVMS described in this document as the management philosophy, process, and system to manage the contract. If the customer does not require EVMS or any Cost Performance Report (CPR), Cost/Schedule Status Report (C/SSR), or equivalent on the contract, the Project Manager may use LGM INTERNATIONAL EVMS in total or selected concepts and disciplines to manage the contract.

1.2 Overview

The LGM INTERNATIONAL EVMS encompasses all methods and disciplines for managing contracts that require more than a minimum amount of cost and schedule status monitoring on a routine basis. The LGM INTERNATIONAL EVMS provides valid, timely, and auditable project cost and schedule status information that relates directly to work progress for all levels of LGM INTERNATIONAL management responsibility. Project Controls is responsible for the maintaining the LGM INTERNATIONAL EVMS to ensure compliance with customer requirements; remaining current on all government guidance with respect to EVMS; maintaining the LGM INTERNATIONAL EVMS description; ensuring and assisting in the implementation of LGM INTERNATIONAL EVMS on contracts; conducting periodic internal surveillance of projects using LGM INTERNATIONAL EVMS to ensure proper implementation, use of the subsystems and disciplines, and the validity of management information; and assisting Project Managers with EVMS and C/SSR reviews of subcontractors and subsequent surveillance of the subcontractors' management systems.

1.3 Content

The LGM INTERNATIONAL EVMS is an integrated management control system comprising a set of major subsystems. This document consists of an overview of the LGM INTERNATIONAL EVMS followed by a description of each major subsystem. It also includes a copy of the Earned Value Management System Evaluation with annotations to provide a cross reference to applicable paragraphs of the system description. This system description plus related company policies, procedures, guidelines, instructions, and project directives comprise the total performance measurement system. This system description is arranged into the following major areas:

- Planning

    - Proposal/Negotiation — Section 3

    - Organization — Section 4

    - Project Budget Planning — Section 5

    - Scheduling — Section 6

    - Work Authorization — Section 7

- Operation/Management

    - Performance Measurement — Section 8

    - Material Planning and Control — Section 9

    - Subcontract Planning and Control — Section 10

□ Accounting

- Cost Accounting — Section 11

- Overhead Cost Management — Section 12

□ Analysis/Reporting/Surveillance

- Data Analysis and Reporting — Section 13

- Revisions and Baseline Control — Section 14

- Surveillance — Section 1

□ Criteria Compliance

- Earned Value Management Implementation Criteria Checklist

1.4   System Objective

The LGM INTERNATIONAL EVMS is designed to fulfill LGM INTERNATIONAL executive and project management and customer management needs and objectives. Effective project management demands a management control system that is responsive to project progress and provides timely information for active management rather than reactive management. LGM INTERNATIONAL active management involves making decisions to minimize the impact of potential problems. The LGM INTERNATIONAL EVMS is not a system of hardware and software, but rather the people using the information to make decisions. LGM INTERNATIONAL EVMS is based on planning and performance data generated at the working level. Data summarized from this level are used by upper management in the decision-making process and provide a sound basis for projecting final costs and future funding requirements.

The LGM INTERNATIONAL EVMS described herein satisfies project management's needs and performs the following functions:

□ Generates contract information in a format suitable for functional management review and analysis

□ Generates contract information in a format that enables "management by exception," while providing the detailed data necessary for in-depth analysis

□ Provides efficient data collection and processing to permit timely detection of potential contract problems

□ Provides rapid response to customer-directed changes and project management decisions

□ Maximizes the use of existing LGM INTERNATIONAL management subsystems without compromising the goals of the overall LGM INTERNATIONAL EVMS

1.5   System Organization

Figure 1-2 portrays the method used to accomplish LGM INTERNATIONAL EVMS functions after contract award or Notice to Proceed:

□ Establishing the contract baseline

□ Authorizing the work, accomplishing the work, and recording associated costs

□ Accomplishing monthly performance measurement and analysis

□ Reporting to upper management and the customer

□ Incorporating baseline revisions

These functions are accomplished in a manner that provides the following:

□ Accountability and traceability from the proposal estimate through the budgeting process to the establishment of the Contract Budget Base (CBB). This includes maintenance and control of the CBB through the revision environment.

□ Accountability and traceability of work from the Statement of Work (SOW) to the lowest level of work authorization.

☐ Clear, concise reporting of contract cost and schedule data from a single, common database to LGM INTERNATIONAL management and the customer.

1.6 Compliance

The LGM INTERNATIONAL EVMS is designed to comply with the EVMS in accordance with DoDI 5000.2R, NASA NPD 9501.3, DOE M 413.3-1, and FAA T 1.13. It is also in line with EVMS Guidelines that provide the basis for determining whether contractors' EVM systems are acceptable and meet the basic requirements of American National Standards Institute/Electronic Industries Alliance (ANSI/EIA) Standard 748-1998 (ANSI/EIA-748).This system description encompasses the policy to be followed by all LGM INTERNATIONAL product lines assigned the responsibility for contracts requiring compliance with EVMS as specified by the customer or as chosen by the LGM INTERNATIONAL COO and/or Director of Project Controls.

1.7 LGM INTERNATIONAL EVMS Internal Surveillance and Surveillance of Subcontractors

The Project Controls of LGM INTERNATIONAL Government Operations is responsible for the maintenance of the LGM INTERNATIONAL EVMS to ensure compliance with customer performance measurement systems requirements. Part of this responsibility includes conducting periodic internal surveillance of programs using LGM INTERNATIONAL EVMS to ensure proper implementation, proper use of the subsystems and disciplines, and the validity of management information. The department also assists programs in the surveillance of subcontractors' management systems for subcontractors with EVMS and C/SSR flowdown from LGM INTERNATIONAL.

Project Controls maintains LGM INTERNATIONAL EVMS surveillance plans, which include review approaches and the LGM INTERNATIONAL organizations that will provide team members for the reviews. LGM INTERNATIONAL EVMS surveillance is structured to ensure that the following major management areas, at minimum, are reviewed at least once per year on the basis of a sampling of programs:

- Organization — Organizational Breakdown Structure, CWBS, CWBS Dictionary, and Responsibility Assignment Matrix (RAM) maintenance

- Project Planning and Budgeting — Scheduling, baseline logs, budget traces to Control Accounts, work authorization documents, material planning and control, subcontract planning and control, proper use of earned value techniques

- Accounting — Estimated Actual Cost of Work Performed (ACWP), subcontractor CPR and C/SSR ACWP, overhead rate application, accounting adjustments, labor charge error correction

- Analysis, Reporting, and Estimate at Completion (EAC) — Variance thresholds, material price and usage variances, labor rate and efficiency variances, variance analysis reports (VARs), estimate to complete (ETC) rationale, EAC update process, comprehensive EACs, CPR and C/SSR accuracy and timeliness

- Revisions — Project change logs, timeliness of change incorporation, rolling wave planning, internal replanning process and controls

The surveillance reviews include a sampling review of logs, data, and forms. The surveillance team interviews a sampling of Control Account Managers (CAMs) to ensure adequate knowledge and use of the LGM INTERNATIONAL EVMS management philosophies, process, and information. If any problems are found, the surveillance team identifies these in a surveillance report to the Project Manager. Serious problems are identified as such, and a corrective action plan is developed by the Project Manager and submitted to Project Controls for approval. These problems are monitored by Project Controls to resolution. Other problems are reviewed during the next surveillance review for determination of closeout of the discrepancy. Copies of all surveillance reports are submitted to the appropriate COO, Product Line Vice President, Director of Project Controls, and Product Line Operations Manager.

Project Controls helps project management conduct surveillance reviews of subcontractors that have either EVMS flowdown with CPR reporting or C/SSR flowdown with C/SSR reporting. Each Project Manager, with flow-down requirements, is responsible for developing a Surveillance of Subcontractor(s) Plan, which uses the LGM INTERNATIONAL EVMS Surveillance Plan as a guide. This plan is submitted to Project Controls to aid in surveillance planning to ensure the availability of sufficient team members to accomplish the reviews.

Use or disclosure of data contained on this sheet is
subject to the restriction on the title page of this document.

Page 4

The basis for the surveillance program is the NDIA program for certification and the surveillance program will make use of spot checks on a monthly basis of the process use by the CAM's.

The use of joint surveillance will be based upon review of the program to ensure that the implementation and maintenance of the EVMS program continues to reflect compliance with the EVMS Criteria. The Joint Surveillance Team (JST) can be used to conduct system surveillance. The JST will establish a surveillance schedule with periodic meetings for the review of EVMS metrics, results from program surveillance activities, results from CPR analysis, results from IBRs and concerns of the Government Program Office. The "Rules of Engagement" document will outline how findings from surveillance will be documented and conflicts resolved.

Data collected through surveillance and open areas of concern from IBRs and Government Program Offices will be used as inputs to the review process. Targeted reviews are conducted by team members (consisting of contractor/Government mix) when surveillance activities point to areas where EVMS compliance indicators are no longer within acceptable limits and other inputs point to areas of concern.

This Joint Surveillance Program will remain in place indefinitely, subject to modification by mutual agreement or termination by either party.

Use or disclosure of data contained on this sheet is subject to the restriction on the title page of this document.

Page 5

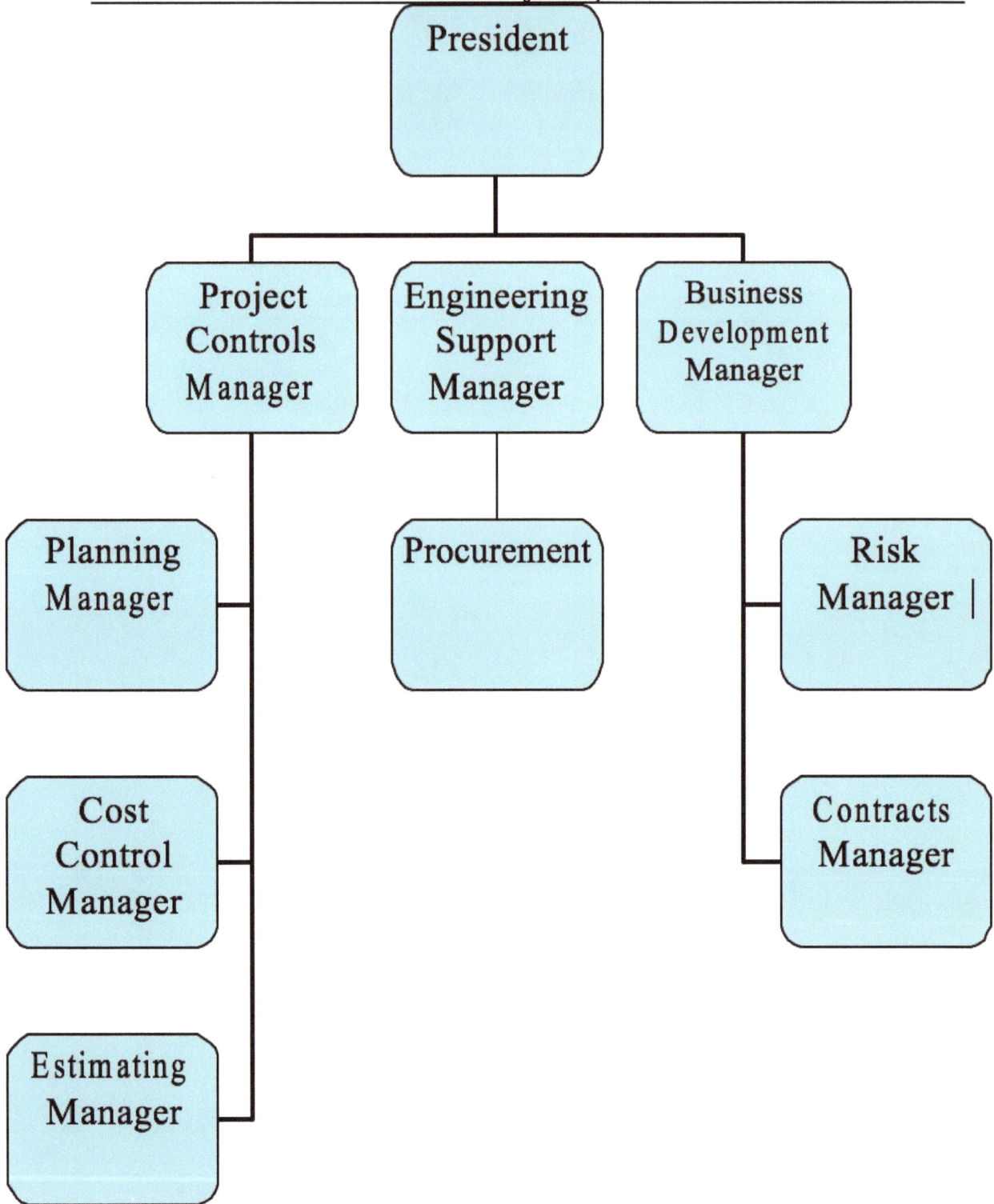

Figure 1-1   LGM INTERNATIONAL Traditional Organizational Chart

Use or disclosure of data contained on this sheet is
subject to the restriction on the title page of this document.

Page 6

CUSTOMER

Award Contract

CPR Format 1-5 → Input Contract Change

BUSINESS DEVELOPMENT/ CONTRACTS

Contract or Authority to Proceed

Contract or Supplemental Agreement (SA)

PROJECT MANAGER/ PROJECT OFFICE

Contract Work Authorization (CWA)

Approve

Approve

Approve

Approve & Sign

Revised CWA

Approve

PROJECT CONTROLS

Establish Preliminary Baseline Schedule

Input/Issue Baseline Schedule

PWA CAMs CAPs

Input to Issue PWA/ BCWS

Input Schedule Status & BCWP

EVMS Data BAse

VAR Form Functional VAR — Updated ETC/EAC & Schedule Forecast

Input ETC/ BAC

EVMS Data BAse

Format 4 / Format 3 / Format 2 / CPR Format 1

Develop CPR Format 5 Problem Analysis

Update Logs, Schedules, Issue Revised PWA(s)

Rev CAP Rev CAWA Revised PWA

Update to Baseline

EVMS Data BAse

FUNCTIONAL MANAGERS

Input to Preliminary Baseline Schedule

Preliminary PWA Budget / Approved Baseline Schedule

CAWA Control Account plan

Final PWA

VAR Form Schedule Control Account Perf Measurement Reports

VAR Form Updated ETC EAC & Schedule Forecast

Prepare Functional VAR if Requested

Preliminary Revised CAWA

CA CAP(s) Revised CAWA

Approve Revised CAWA & CAP(s)

CONTROL ACCOUNT MANAGERS

Input to Preliminary Baseline Schedule

Prelim CAWA / Approved Baseline Schedule

Final CAWA

Start Work

Determine Progress/ Accomplishment

Estimated ACWP

Schedules VAR Form Control Account Perf Measurement Rpts by Wk Pkg

Schedules CPR or C/ SSR

Analyze Variance Write VAR & Update ETC/EAC & Schedule Forecast

Determine if CAP Change Required

Prepare Revised CAP & Prelim Revised CAWA

Preliminary Revised CAWA

Approve Revised CAWA & CAP(s)

Perform Work

FINANCE/ACCOUNTING

Direct Labor & ODC ACWP

PROCUREMENT MATERIALS

Purchase Requisition

Prepare Priced Issue Report & MMR

SUBCONTRACTORS

Purchase Order or Subcontract

Start Work

CPR C/SSR or Progress Report

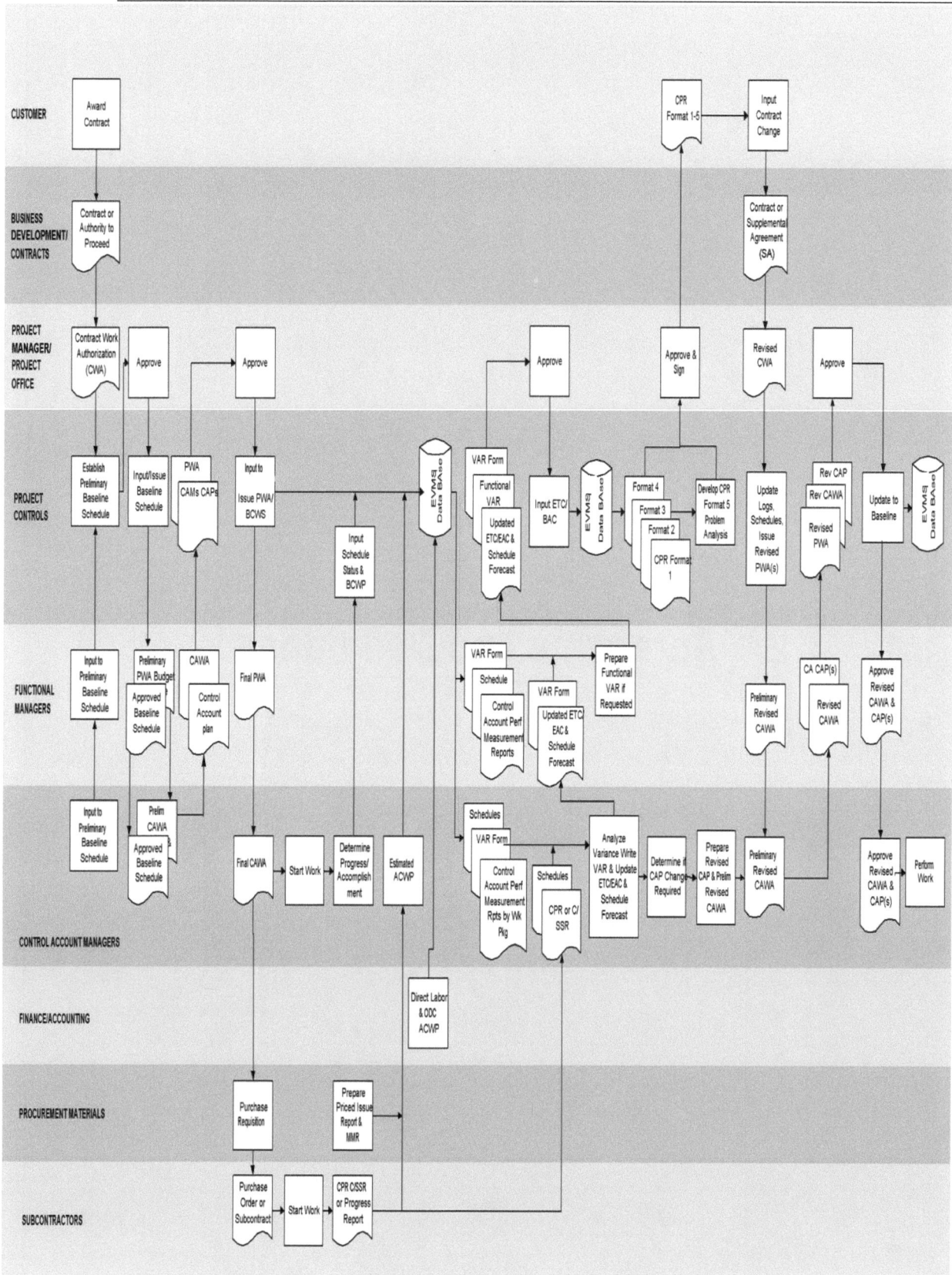

Figure 1-2   LGM INTERNATIONAL EVMS Overview Flowchart

Use or disclosure of data contained on this sheet is subject to the restriction on the title page of this document.

Page 7

2. Acronyms and Definitions

Use or disclosure of data contained on this sheet is
subject to the restriction on the title page of this document.

Page 8

2.1 Acronyms and Abbreviations

Acronym	Meaning
AA	Advance Agreement
ACO	Administrative Contracting Officer
ACWP (AC)	Actual Cost of Work Performed (Actual Cost)
AUW	Authorized Un-Priced Work
BAC	Budget at Completion
BCWP (EV)	Budgeted Cost for Work Performed (Earned Value)
BCWS (PV)	Budgeted Cost for Work Scheduled (Planned Value)
LGM EVMS	LGM INTERNATIONAL Earned Value Management System
CA	Control Account
CAM	Control Account Manager
CAP	Control Account Plan
CAS	Cost Accounting Standard
CAWA	Control Account Work Authorization
CBB	Contract Budget Base
CCDR	Contractor Cost Data Report
CDRL	Contract Data Requirements List
CEI	Contract End Item
CFSR	Contract Funds Status Report
CLIN	Contract Line Item Number
CPI	Cost Performance Index
CPR	Cost Performance Report
C/SSR	Cost/Schedule Status Report
CTC	Contract Target Cost
CTP	Contract Target Price
CV	Cost Variance
CWA	Contract Work Authorization
CWBS	Contract Work Breakdown Structure
DB	Distributed Budget
DCAA	Defense Contract Audit Agency
DCMA	Defense Contract Management Agency
DPRO	Defense Plant Representative Office
DoD	U.S. Department of Defense
EAC	Estimate at Completion
ETC	Estimate-to-Complete
EVMS	Earned Value Management System
FAR	Federal Acquisition Regulation
FM	Functional Manager
G&A	General and Administrative
IBR	Integrated Baseline Review
LGM	LGM International
LOE	Level of Effort
MIL-STD	Military Standard
MOA	Memorandum of Agreement
MOU	Memorandum of Understanding

Acronym	Meaning
MR	Management Reserve
MRR	Material Receiving Report
OBS	Organizational Breakdown Structure
ODC	Other Direct Cost
OTB	Over Target Baseline
PIR	Perpetual Inventory Record
PM	Project Manager
PMB	Performance Measurement Baseline
PMS	Project Master Schedule
PO	Purchase Order
PC	Project Controls
PV	Price Variance
PWA	Project Work Authorization
RAM	Responsibility Assignment Matrix
RFP	Request for Proposal
SDRL	Subcontractor Data Requirements List
SOW	Statement of Work
SPI	Schedule Performance Index
SV	Schedule Variance
TAB	Total Allocated Budget
TCPI	To Complete Performance Index
UB	Undistributed Budget
UV	Usage Variance
VAC	Variance at Completion
VAR	Variance Analysis Report
WP	Work Package

The utilization of the Abbreviations are based on the original standards, the utilization of PV, AC, EV can be interchanged, but for process consistency it will maintain BCWS/ACWP/BCWP based on traditional process.

2.2   LGM INTERNATIONAL Earned Value Management System Glossary

Activity. A time-consuming element in the execution of a task. Activities start upon the completion of all preceding activities, and they must be completed before succeeding activities can start. Usually represented on a network by a solid line between two events.

Actual Cost. See Actual Cost of Work Performed.

Actual Cost of Work Performed (ACWP). The dollar value of resources consumed in the accomplishment of work performed. This includes actual direct costs such as the incurred costs of labor, material, subcontract, and Other Direct Costs plus the related overhead/General and Administrative costs.

Actual Direct Costs. Costs identified specifically with a contract, based on the contractor's cost identification and accumulation system as accepted by cognizant Defense Contract Audit Agency representatives. See Direct Costs.

Advance Agreement. An agreement between the contractor and the Contract Administration Office concerning the application of an approved EVMS to contracts within the affected facility.

Allocated Budget. See Total Allocated Budget.

Applied Direct Costs. The actual direct costs recognized in the time period associated with the consumption of labor, material, subcontracts, and Other Direct Costs, regardless of their date of commitment or date of payment. These amounts are charged to work in process when any of the following takes place:

☐ Labor, material, or other direct resources are actually consumed

☐ Material resources are withdrawn from inventory for use

☐ Material resources are received that are uniquely identified to the contract and scheduled for use within 60 days

☐ Cost Performance Report or Cost/Schedule Status Report is received from subcontractor or earned value is established by the CAM

Apportioned Effort. Effort that is not readily divisible into short-span work packages but that is related in direct proportion to measured effort.

Authorized Un-Priced Work. Work authorized by the customer that has not yet been negotiated and definitized.

Authorized Work. Definitized, on-contract effort and effort for which contract costs have not been agreed to but for which written authorization has been received.

Baseline. See Performance Measurement Baseline.

Bid and Proposal Costs. Costs incurred in preparing, submitting, and supporting bids and proposals (whether solicited or not) on potential government and non-government contracts.

Budget at Completion. The sum of all budgets established for the contract. See Total Allocated Budget.

Budgeted Cost for Work Performed (or Earned Value). The sum of the budgets for completed work packages and completed portions of open work packages, plus the applicable portion of the budgets for Level of Effort and Apportioned Effort.

Budgeted Cost for Work Scheduled (BCWS) (or Planned Value). The sum of the budgets for all work packages, planning packages, etc., scheduled to be accomplished (including in-process work packages), plus the amount of Level of Effort and Apportioned Effort scheduled to be accomplished within a given time period.

Burden. Overhead expenses (labor, material, and other) not directly chargeable to a specific job order and, therefore, distributed over the appropriate direct labor and/or material base.

Change Order. A formal authorization by Procuring Contracting Officers for a change or variance to an existing contract. See Supplemental Agreement.

Commitment. The incurrence of a liability for goods or services; the portion of the goods or services that have been ordered but no payment has been made. The point in which a contractor enters into an agreement with a supplier, generally when a purchase order is transmitted; the point of original commitment.

Contract Budget Base. The negotiated contract cost plus the estimated cost of authorized un-priced work.

Contract Data Requirement List. A compilation of all data requirements the contractor is obligated to submit to the government.

Contract End Item. Items specified for contract delivery.

Contract Funds Status Report. DoD report that provides information to accomplish the following:

☐ Update and forecast contract fund requirements

☐ Plan and decide on funding changes

☐ Develop fund requirements and budget estimates in support of approved programs

Contract Line Item Number. Supplies and services that appear as itemized entries in the contract.

Contractor Cost Data Report. Identifies the contractually required customer report defined in DoD 5000.2M. This report is developed to give DoD components a means to collect contract cost and related data to aid in acquisition management. It is designed to collect data on defense material items in a standard format to conduct cost estimating, programming, budgeting, and procurement responsibilities.

Contract Target Cost. The dollar value (excluding fee or profit) negotiated in the original contract plus the cumulative cost (excluding fee or profit) applicable to all definitized changes to the contract. It consists of the estimated cost negotiated for a cost-plus-fixed-fee contract and the definitized target cost for an incentive contract. The Contract Target Cost does not include the value of authorized/un-negotiated work and is thus equal to the Contract Budget Base only when all authorized work has been negotiated and definitized.

Contract Target Price. The negotiated cost plus planned profit or fee.

Contract Work Authorization. An internal document that authorizes the Project Manager to accomplish contract work.

Contract Work Breakdown Structure. The complete work breakdown structure for a contract, including the DoD-approved work breakdown structure for reporting purposes and its discretionary extension to the lower levels by the contractor, in accordance with MIL-HNDBK 881 (latest version).

Control Account. A management control point at which budgets, earned value, estimates, and actual costs are accumulated and compared for variance analysis. A natural control point for cost/schedule planning and control formed by the intersection of the organizational, Contract Work Breakdown Structure (CWBS), and element of cost dimensions. The Control Account is a natural control point for cost/schedule planning and control because it represents the work assigned to one responsible organization on one CWBS element.

Control Account Manager. The individual responsible for the management of a Control Account. The Control Account Manager is responsible for planning and managing the resources assigned to accomplish the task.

Control Account Plan. The detail plan prepared by the Control Account Manager showing time-phased planning of Work Packages and Planning Packages and their associated budget for a Control Account. This plan can be documented/recorded through the use of a Control Account planning form or an equivalent combination of forms and/or reports. The information must be broken down by Work Package and Planning Package and Budgeted Cost for Work Scheduled by month by element of cost, scope of work, schedule start and stop, and earned value method.

Control Account Work Authorization. A document that authorizes the Control Account Manager to accomplish a specified scope of work within an identified time frame for an agreed-on budget.

Cost of Money. Costs associated with the time use of money as defined in the contract.

Cost Performance Index (CPI). The value earned for every unit of actual cost expended.

CPI = BCWP/ACWP   CPI Greater Than 1.0 = Underrun
                       CPI Equal to 1.0 = On Target
                       CPI Less Than 1.0 = Overrun

Cost Performance Report. A contractually required report, prepared by the contractor, containing information derived from the internal EVMS. Provides status of progress on the contract.

Cost/Schedule Status Report. A performance measurement report.

Cost Variance. A metric for the cost performance on a contractor program. It is the algebraic difference between earned value and actual cost (Cost Variance = Earned Value – Actual Cost). The dollar value of work accomplished for each dollar spent. A positive value indicates a favorable position and a negative value indicates an unfavorable condition.

Critical Path Analysis. See Network Schedule.

Defense Contract Audit Agency. The organization tasked with monitoring a contractor's design and implementation of an accounting system.

Defense Contract Management Agency. Government offices located at numerous contractor facilities throughout the United States. Their primary function is to administer all contract administrative activities.

Detailed Schedule. Detailed schedule of work packages that make up a Control Account.

Direct Costs. Costs that may be identified with a particular cost objective.

Direct Labor. The portion of labor expended in designing, tooling, testing, and physically applying labor to material, altering its shape, form, or nature, or in fulfilling a contractual requirement for service.

Discrete Effort. Tasks related to the completion of specific end products or services that can be directly planned and measured.

Discrete Milestone. A milestone with a finite, scheduled occurrence in time, signaling the finish of an activity (e.g., "release drawings," "pipe inspection complete") and/or signaling the start of a new activity.

Distributed Budget. Budget applicable to contract effort that has been identified to Contract Work Breakdown Structure elements at or below the lowest level of reporting to the government.

Earned Value (or Budgeted Cost for Work Performed). The value of completed work expressed in terms of the budget assigned to that work.

Earned Value Management System (EVMS). An integrated management system that uses earned value to measure progress objectively.

Element of Cost. A category or type of cost such as labor, material, subcontracts, and Other Direct Costs. Separate codes are established for each major element of cost to aid in the accounting and control processes.

Estimate at Completion. Actual direct costs, plus indirect costs allocable/allowable to the contract, plus the estimated costs (direct and indirect) for authorized work remaining.

Estimate to Complete. The portion of the EAC that addresses total expected costs for all work remaining on the contract.

Formal Reprogramming. Restructuring of the effort remaining in the contract that results in a new project budget allocation that exceeds the Contract Budget Base. The process may eliminate cumulative schedule and cost variances and produce a new Performance Measurement Baseline. The customer must be consulted in advance whenever an Over Target Baseline is implemented on a government contract.

Functional Manager. The manager of a significant portion of the project effort within the Organizational Breakdown Structure, typically responsible for like effort (e.g., engineering, construction, operation).

Functional Organization. An organization or group of organizations with a common operational orientation (e.g., engineering, construction, operations).

Gantt Chart. A graphic representation used as an aid to effective scheduling and control by setting up graphically on a time scale when certain events are to take place or where deadlines occur; a sophisticated bar chart.

General and Administrative. Expenses incurred in the direction, control, and administration of the company (including selling expenses). These expenses are spread over the total direct and burden cost at a negotiated rate.

Independent Estimate at Completion. A statistically computed forecast based on performance to date and a mathematical projection of this performance to derive the estimated contract cost at completion.

Use or disclosure of data contained on this sheet is subject to the restriction on the title page of this document.

Page 13

Indirect Cost. A cost that, because of its incurrence for common or joint objectives, is not readily subject to treatment as a direct cost.

Integrated Baseline Review. A joint contractor and government review, typically conducted within 180 days after contract award, used to assess the contractor's planning, Statement of Work, schedule logic, resource adequacy, and risk assessment.

Intermediate Schedule. A schedule that illustrates a level of detail between the Project Master Schedule and the Detailed Schedule.

Internal Replanning. Actions performed by the contractor for remaining scope within the budget and schedule constraints of the contract. The contractor is required to notify the government of all internal replanning actions, but the government has no approval authority over this action.

LGM INTERNATIONAL Earned Value Management System (LGM INTERNATIONAL EVMS). LGM INTERNATIONAL's system for managing programs, projects, and contracts.

Labor Rate Variances. A component of a Labor Cost Variance. Difference between planned labor rates and actual labor rates. Labor rate variances are derived by (BCWP Rate less ACWP Rate) x ACWP Hours.

Level of Effort. Effort of a general or supportive nature that does not produce definite end products.

Make or Buy. The identification of major hardware components of a contract about whether they will be fabricated by LGM INTERNATIONAL (Make) or obtained from outside sources (Buy).

Management Reserve. An amount of the Total Allocated Budget withheld for management control purposes rather than designated for accomplishing a specific task or set of tasks. It is not a part of the Performance Measurement Baseline.

Material Receiving Report. A document in which material receipts are recorded. This report is used in the vendor payment–approval process, in updating material inventory records, and by the Control Account Manager to determine Budgeted Cost for Work Performed for material received.

Measured Effort. Effort that is measured discretely. This definition is used to separate effort measured by Level of Effort and generated Apportioned Effort.

Memorandum of Agreement (MOA). A document that establishes agreement between the cognizant onsite government agency and the LGM INTERNATIONAL Project Office to ensure a complete LGM INTERNATIONAL EVMS surveillance project. The MOA delineates the responsibilities of the procuring agency and the cognizant government office and addresses surveillance activities, frequency of audits and reports, depth and detail of analysis, and notification of deficiencies and other special problems unique to the contract.

Milestones. Finite defined events that constitute start or completion of a task or occurrence of an objective criterion for accomplishment. Milestones should be discretely measurable; the passage of time itself is not sufficient. However, milestones should be associated with a schedule date to determine when the milestone is to occur.

Network Schedule. A schedule format in which the activities and milestones are represented along with the interdependencies between activities. It expresses the logic of how the program will be accomplished. Network schedules are the basis for critical path analysis, a method for identification and assessment of schedule priorities and impacts.

Non-recurring Costs. Expenditures against specific tasks that do not occur on a repetitive basis in any given project (e.g., expenditures for preliminary design effort, qualification testing, and initial tooling and planning).

Organizational Breakdown Structure. A functionally oriented division of the contractor's organization established to perform the work on a specific contract.

Original Budget. The budget established at, or near, the time the contract was signed, based on the negotiated contract cost.

Other Direct Costs. Costs that can be isolated to specific tasks other than labor and material (e.g., travel, computer time, services).

Overhead Budget. The budget value established for costs to be incurred by persons and/or departments for tasks that do not have a direct relationship to the end cost objective.

Over Target Baseline. Baseline that results from reprogramming.

Performance Measurement Baseline. The time-phased budget plan against which contract performance is measured. It is formed by the budgets assigned to Control Accounts and applicable overhead budgets. For future effort not planned to the Control Account level, the performance measurement baseline includes budgets assigned to higher level Contract Work Breakdown Structure elements and Undistributed Budgets. It equals the Total Allocated Budget less Management Reserve.

Performance Period. A time interval of contract performance.

Performing Organization. A defined unit within the contractor's organizational structure that applies the resources to perform the work.

Planning Package. A logical aggregation of downstream work within a Control Account, usually the downstream effort, that can be identified and budgeted in early baseline planning but cannot be defined into work packages yet.

Price Variance. Component of a Material Cost Variance. Difference between the planned cost of a purchase item and its actual cost. Price variance is derived by (BCWP Unit Price – ACWP Unit Price) x ACWP Quantity.

Procuring Activity. The subordinate command to which the Procuring Contracting Officer is assigned. It may include the project office, related functional support offices, and procurement offices.

Product Line. A profit center entity within the LGM INTERNATIONAL corporate organization.

Project Directive. A document giving specific contract operation instructions. A Project Directive may be issued to inform functional organizations of project requirements, selected control system options, and project execution responsibilities.

Project Manager. A generic term used to refer to the person responsible for executing the overall planning, direction, control, coordination, and evaluation of the assigned project. Variously designated as Project Manager, Project Leader, or Project Director.

Project Master Schedule. The highest summary level schedule for a project depicting overall project phasing and interfaces, contractual milestones, and project elements segregated by functional responsibility in support of specific project objectives.

Project Management Office. The office of the Project Manager, typically including the Project Manager, Project Controls, and Project Finance.

Project Risk Analysis. System that provides a continuous analysis of identified risks with respect to their impact on project cost, schedule, and technical performance.

Project Work Authorization. A document that authorizes a major functional organization to accomplish a specified scope of work within an identified schedule for an agreed-on budget.

Purchase Order. A written agreement of any transaction between LGM INTERNATIONAL and a supplier in which materials, parts, facilities, or services are exchanged for financial consideration. A Purchase Order involves a contract, the mutual understanding and obligations of which must be a matter of record.

Recurring Costs. Expenditures against specific tasks that occur on a repetitive basis (e.g., sustaining engineering, production of operational equipment, tool maintenance).

Replanning. The redistribution of budget for future work. Traceability to previous baselines is required and attention to funding requirements must be considered in any replanning effort.

Reporting Level. A level or levels of the Contract Work Breakdown Structure and Organizational Breakdown Structure designated for formal performance reporting to the government.

Responsibility Assignment Matrix. A depiction of the relationship between Contract Work Breakdown Structure elements and organizations assigned responsibility for ensuring their accomplishment.

Responsible Organization. A defined unit within the contractor's organizational structure that is responsible for accomplishing specific tasks.

Rolling Wave Planning. The progressive refinement of detailed work definition by continuous subdivision of downstream activities into near-term tasks or the conversion of Planning Package tasks into Work Packages.

Schedule Indices (Schedule Index). A performance indicator reflecting the relationship of earned value to budget.

Schedule indices = BCWP/BCWS   Over 1.0 = ahead of schedule
                                  Equal to 1.0 = on schedule
                                  Under 1.0 = behind schedule

Schedule Variance. A metric for the schedule performance on a project. Schedule variance is the algebraic difference between earned value and the budget (Schedule Variance = Earned Value – Budget). The dollar value of work accomplished for each dollar of work performed. A positive value is a favorable condition; a negative value is unfavorable.

Significant Variances. Differences between planned and actual performance that exceed pre-established variance thresholds and require further review, analysis, or action.

Statement of Work. Document that defines the work scope requirements of a project.

Subcontract. Includes the cost of labor, materials, and/or services performed by vendors other than LGM INTERNATIONAL in accordance with LGM INTERNATIONAL's Statement of Work.

Target Cost. See Contract Budget Base and Contract Target Cost.

To-Complete Performance Index. Cost efficiency that must be achieved in the remaining period of performance to complete the total work scope within the budget (Budget at Completion) or cost (Estimate at Completion) objective.

Total Allocated Budget. The sum of all budgets allocated to the contract. The Total Allocated Budget consists of the performance measurement baseline and all Management Reserve and reconciles directly to the Contract Budget Base. Any difference in quantity and cause is documented.

Undistributed Budget. Budget applicable to contract effort that has not been identified yet to Contract Work Breakdown Structure elements at or below the lowest level of reporting to the government.

Usage Variance. A component of Material Cost Variance. Usage variance is the difference between earned quantity of materials and actual quantity used, expressed in dollars and is derived by (BCWP Quantity – ACWP Quantity) x BCWP Unit Price.

Variance. The value by which any schedule or cost performance varies from a specific plan. Significant variances are differences between planned and actual performance, which require further review, analysis, or action. Appropriate thresholds are established for variance analysis. See Cost Variance, Schedule Variance, and Cost Variance at Completion.

Variance Analysis Report. A report describing the nature, cause, impact, and corrective action for variances that exceed agreed-on thresholds.

Variance at Completion. The difference between the total budget assigned to a contract, Contract Work Breakdown Structure element organizational entity, or Control Account and the Estimate at Completion. Variance at Completion = Budget at Completion – Estimate at Completion. It represents the amount of expected overrun or underrun.

Variance Threshold. Internal and external tolerances or thresholds established by management direction or the customer. Variance conditions outside the threshold limits require investigation, analysis, reporting, and corrective action.

Work Breakdown Structure. A WBS is the result of project/program planning that establishes the physical work packages or elements and the activities within those packages that completely define a project.

Work Package. Detailed jobs, or material items, identified by the contractor for accomplishing work required to complete the contract. A work package has the following characteristics:

- It represents units of work at levels where work is performed.
- It is clearly distinguished from all other work packages.
- It is assigned to a single organizational element.
- It has scheduled start and completion dates and, as applicable, interim milestones that represent physical accomplishment.
- It has a budget or assigned value expressed in terms of dollars, labor hours, or other measurable units.
- Its duration, or its level of effort (LOE), is limited.
- It is integrated with detailed engineering, construction, or other schedules.

Use or disclosure of data contained on this sheet is subject to the restriction on the title page of this document.

Page 17

3. Proposal/Negotiation

Use or disclosure of data contained on this sheet is
subject to the restriction on the title page of this document.

Page 18

3.1　Overview

At LGM INTERNATIONAL, EVMS principles are applied to a project beginning with the proposal. Project cost/schedule considerations are essential in the proposal preparation process because during this phase:

☐ The Contract Work Breakdown Structure (CWBS) is usually either extracted from the Request for Proposal (RFP) CWBS or is developed.

☐ CWBS elements are used in the proposal process to build the schedule and estimate/cost proposed price.

☐ The Organizational Breakdown Structure (OBS) is defined and extended to the level of proposed management responsibility.

☐ The scope of work is delineated.

☐ The CAMs are identified by responsible organization within the OBS.

The structure established during the proposal stage and altered as necessary during negotiations serves as the basis of the Performance Measurement Baseline. Although initial resource estimating is done during the proposal phase, after contract award, there may be a need to update the resource estimate to include such things as changes to the schedule, budget, or scope of work; establishment of a Management Reserve (MR); or adjustments required to accommodate resource availability.

3.2　Establishing the Proposal Plan

3.2.1　Initial Planning

Initial planning is conducted during the proposal phase and continues through award and definitization. During the initial phase, customer requirements are translated into meaningful plans for development. As stated in the overview, several key elements of the LGM INTERNATIONAL EVMS are used to support the proposal and become the basis of system documentation after contract award. Figure 3-1 illustrates the general process flow of establishing the proposal baseline. Due to the nature of programs and projects LGM INTERNATIONAL performs, the Proposal Department maintains a proposal process flow conforming to the principles established in Figure 3-1 and maintains procedures detailing specific approval/review limits and related details. Figure 3-1 also shows the preparation and review cycle of the supporting proposal documentation from the RFP to submission of the proposal to the customer.

3.2.2　Critical Proposal Elements

A smooth transition from proposal to contract is critical to a successful contract, and consideration must be given to critical planning elements. A bid decision should initiate the review or development of the following:

☐ Bidding Instructions, SOW, Contract Work Breakdown Structure (CWBS), and CWBS Dictionary

☐ OBS

☐ Major Milestone Planning

Each of these planning activities is discussed in the following paragraphs.

3.2.2.1　Bidding Instructions

Bidding instructions may have a significant impact on the cost of the effort. Copies are made available to all parties at the proposal kickoff meeting. They usually highlight the key requirements of the RFP, identify organizational responsibilities, and provide other relevant information. Initial make or buy decision planning frequently begins at this point. When responding to public sector RFPs (international, state, or local governments), the senior contracts official within the product line will review the entire bid document and comment on the bidding instructions. On federal procurements, the Government Contracts Department within LGM INTERNATIONAL will review and comment on the RFP.

3.2.2.2 Statement of Work

LGM INTERNATIONAL EVMS planning relates the RFP SOW to the proposal's detailed plans for accomplishment. Upon contract award these detailed plans are expanded based on the negotiated SOW. A further expansion of the negotiated SOW is contained in the CWBS Dictionary, which describes each element of the work.

3.2.2.3 Contract Work Breakdown Structure

LGM INTERNATIONAL uses a CWBS on all major contracts, regardless of whether it is required by the customer. The CWBS is the common framework for defining the elements of authorized work to meet requirements of the contract. Whether the CWBS is provided in the RFP or developed by LGM INTERNATIONAL, accountability and traceability from the proposal to the contract require a CWBS that is structured to meet the following minimum requirements:

- It is prepared using customer guidance documents as appropriate; for example, DoD MIL-STD 881 (latest revision), Department of Energy MA-0295, and National Aeronautics and Space Administration Handbooks NASA-NHB 9501.2B and NHB 5610.1.
- Contract line items and end items are identified.
- It is structured to contain all work elements of the contract.
- Contains elements identifying selected subcontractor effort with subcontractor name in parenthesis
- It clearly relates to the SOW.
- It is extended to the lowest level at which organizational responsibility will be identified.

The CWBS Dictionary is developed at the lowest extended CWBS level to ensure that all work described in the SOW is included in the WBS.

3.2.2.4 Organizational Breakdown Structure

An OBS is developed as soon as the basic organizational structure is defined. The OBS serves several purposes during the proposal stage and after contract award:

- It provides for ease of reference in the text by proposal writers.
- It provides a structure for the Basis of Estimate for the technical/management and cost proposals.
- It provides a structure on which to relate the organization to the CWBS.
- It provides a tool to tie the Control Accounts (CAs) to CAMs.

3.2.2.5 Milestone Planning (Schedule)

Initial milestone planning requires a detailed effort by the proposal team. Care is exercised so that no significant or critical event is overlooked in the definition of major milestones. The hierarchy of milestones that form the basis of the schedule baseline is determined as follows:

- Identification of all items, activities, and events that require specific customer review and acceptance. This is based on an examination of customer directives, instructions, and RFP guidelines. Careful attention is paid to Contract Data Requirements List (CDRL) items.
- Identification of critical tasks (including subcontracts and long lead items) that must be completed for achievement of contract objectives. They are selected from a detailed review of the SOW/CWBS.
- Identification of other activities and events necessary for achievement of contract objectives (e.g., design reviews, configuration management, baseline configuration establishment, make or buy decisions).

Through an evaluation of these milestones, qualifying major milestones are determined for the proposed Project Master Schedule (PMS). Major milestones are the most significant events in the contract cycle. Other milestones are placed in Intermediate Schedules and form the basis of baseline schedule planning at a more detailed level after contract award.

3.3   Proposal/Negotiation Process

Figure 3-1 shows the major activities that take place during the proposal/negotiation process. The following describes this process.

3.3.1   Business Management Decision Process

A. Business Development/Contracts collect and analyze customer data, including the RFP. Business Development/Contracts has the primary responsibility for gathering and analyzing data for procurements.

B. Based on customer data, Business Development/Contracts interprets the customer requirements. Business Development/Contracts estimates the potential value of the contract, sets company objectives in light of the requirement and contract value, and sets the company's overall bid strategy. These data are formatted and presented to the product line (and corporate, if appropriate) management in a bid/no-bid meeting.

C. LGM INTERNATIONAL Corporate Management conducts a meeting to decide whether to bid on a specific RFP.

D. The bid/no-bid decision is made.

3.3.2   Establishing a Proposal Team

E. If the company decides to bid, a Proposal Manager is named by Business Development. The proposal team typically includes the Proposal Manager, the proposed Project Manager (if available), and representatives from Operations, Project Controls, Procurement, and Business Development/Contracts.

F. The Proposal Manager defines the proposal budget and authorizes, through an accounting-approved charge number, proposal expenditures.

G. Based on the overall strategy defined earlier, the proposal team defines in more detail the specific strategies for the technical, management, and cost proposals.

3.3.3   Proposal Definition

H. Based on the detailed proposal strategy, the team defines the technical, management, and costing approaches in more detail.

I. In defining the project in detail, the proposal team interrelates with functional organization elements as necessary. This interrelation constitutes mini-negotiations involving technical approach, work scope, schedule, and resources required.

J. Once approach, scope, schedule, and resources agreements have been reached, the proposal team details these by expanding the SOW/technical verbiage and defines proposal labor-hour requirements.

K. The proposal team assembles a proposed project plan based on inputs from all functional elements. The proposed project plan includes a detailed SOW reflecting RFP requirements and bid instructions, a single CWBS, an OBS, a schedule, a labor-hour basis of estimate, identification of material, and identification of subcontractors.

L. With the proposed project plan in place, the proposal team coordinates the estimating/costing effort.

3.3.4   Estimating the Proposal

M. In developing the estimate, the responsible member of the proposal team derives the labor estimates (in time units, not dollars) for a given task usually using Estimating and Pricing. The method of estimating falls into one of the following categories:

1. Order of Magnitude (t50%)

    -   Prepared by using cost capacity curves or using plant cost ratios

2. Factored Estimate (t25%)

    -   Prepared using factors, quantity ratios, or modeling techniques

3. Control Estimates (t15%)

Use or disclosure of data contained on this sheet is
subject to the restriction on the title page of this document.

Page 21

- Prepared using detailed conceptual estimating techniques from preliminary design information

4. Definitive Estimate (t5%)

   - Prepared using completed design documentation with finalized quantity information from drawings or other documents

5. Design, Technical, Engineering, and Operations Estimates

   - Prepared using detailed labor hours for the basic, detailed, and other project controls associated with a project

6. Maintenance Estimates

   - Prepared using Technique 3 above

O. Estimating and Pricing applies the appropriate labor rates to the labor-hour requirements. Rates may be local customary (obtained by onsite survey), Davis Bacon, Service Contract Act, or LGM INTERNATIONAL standard. For Davis Bacon or Service Contract Act rates, Business Development/Contracts approves the rates and labor burdens.

P. Material take-offs or estimates are developed and subcontract work is quantified by the responsible member of the proposal team usually using the Estimating and Pricing.

Q. Material and subcontracts work is packaged for pricing and/or quotation.

R. Procurement typically sources material, obtaining supplier quotes and schedules, and obtains quotes from subcontractors.

S. Other direct costs (ODCs) (e.g., travel, administrative) are identified by the responsible proposal team member.

Non-labor costs are based on quotes from suppliers or subcontractors whenever possible. Other acceptable sources of costs are as follows:

- Historical records
- Published price sheets
- Catalogs
- Oral quotes
- Engineering estimates

Estimates are documented for the cost file and include the basis of estimate in sufficient detail to justify the estimate and provide meaningful data to the contract budgeting effort.

Proposed subcontractor quotes are selected based on ability to perform, technical capability, understanding of the work scope, schedule compliance, and price.

3.3.5 Proposal Review

T. The cost proposal cost team summarizes the estimates to develop the proposed baseline costs for the project. This summary is prepared in accordance with the bid instructions, in a way that clearly displays costs by appropriate summaries for review. The summary is built up by CWBS or Contract Line Item Number. Business Development/Contracts reviews the proposed costs for customer requirements and company objectives and strategy.

U. Accounting determines the appropriate overhead and General and Administrative (G&A) rates. Bid G&A rates for proposals for federal government procurements are approved by government compliance.

V. Business Development/Contracts reviews the proposal and generates an initial profit rate.

W. LGM INTERNATIONAL Corporate Management reviews the proposed costs and selling prices and provides redirection or recommendations.

X. The proposal is submitted to the client.

Use or disclosure of data contained on this sheet is subject to the restriction on the title page of this document.

Page 22

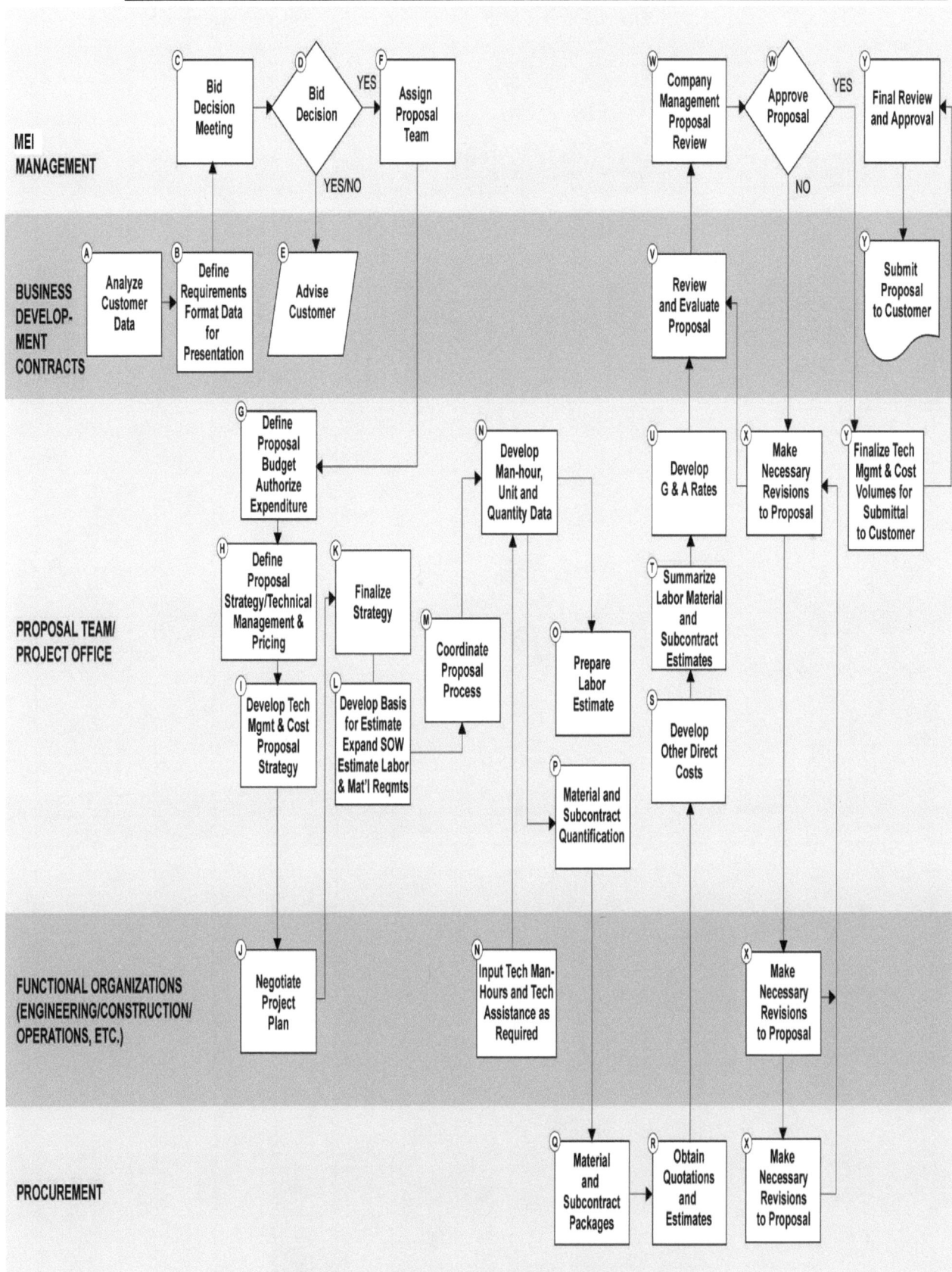

Figure 3-1   Proposal/Negotiation Flowchart

Use or disclosure of data contained on this sheet is
subject to the restriction on the title page of this document.

Page 23

4. Organization

Use or disclosure of data contained on this sheet is
subject to the restriction on the title page of this document.

Page 24

4.1   Overview

The Organization section of the LGM INTERNATIONAL EVMS defines the work to be performed and addresses the assignment of tasks to the functional organizations responsible for performing the work. The organization process includes the following.

- Contract Work Breakdown Structure (CWBS) — Defines all authorized work to meet the requirements of the contract.

- Organizational Breakdown Structure (OBS) — Identifies the functional organizations responsible for accomplishing the authorized work.

- Responsibility Assignment Matrix (RAM) — Integrates the CWBS and the OBS to identify Control Accounts.

- Control Accounts and Work Packages — Control Accounts are subdivisions of the lowest level CWBS element. Work Packages are subdivisions of a Control Account used by the CAM to assist in managing the work and monitoring performance.

Figure 4-1 illustrates the CWBS and RAM development flow.

4.2   Contract Work Breakdown Structure

4.2.1   Identification of Contract Work Breakdown Structure

The CWBS (Figure 4-2) identifies all project work from the contract level to the lower levels of manageable Control Accounts. It is the basis for organizing the work and measuring performance. The CWBS provides the product-oriented family tree of hardware, software, services, and other work tasks and organizes, defines, and graphically displays the product to be produced as well as the work to be accomplished to achieve the specified product. This product-oriented breakdown allows summarization and analysis by end product regardless of functional organizations working on the end products.

4.2.1.1   Definition of Work

Project definition is a top-down activity that considers contract end items, associated contractual delivery schedules and specifications, the contract SOW, and CBB. Project definition is normally accomplished following receipt of a signed contract, although a substantial part of the effort is accomplished as part of proposal preparation and negotiation.

4.2.1.2   Contract Work Breakdown Structure

The CWBS reflects all contract work to be done and the way it is to be managed and performed. CWBS development is a LGM INTERNATIONAL mandatory requirement on all LGM INTERNATIONAL programs because all authorized work is defined within the framework of the CWBS. A single CWBS is used on a contract. The basic objective of the CWBS is to subdivide the total contract into manageable units of work. The CWBS contains all contractually authorized requirements.

The CWBS serves many purposes and facilitates planning by providing a formal structure for identifying the work. It simplifies the problems of summarizing contract data and serves as the reporting structure for customer-required data submittals.

The CWBS is prepared for each contract to reflect the contract SOW, including contract line items, CWBS elements specified for external reporting, CWBS elements to be subcontracted, and the Control Account level. The CWBS, which reflects all contractual work, including all subcontracted work, is developed by extending the summary level WBS provided in the contract. Following are some examples of the guidance documents used in extending the summary WBS down to the level of management control, depending on the customer and/or the type of contract: DoD MIL-STD 881-B, Department of Energy G 430.1-1, and National Aeronautics and Space Administration NHB 7120.5 Major subcontractors are identified in the CWBS block diagram or the CWBS Index. The level of detail in a subcontractor's CWBS is independent of the level of detail in the prime contract CWBS. There is only one CWBS for any given contract.

4.2.2 CWBS Development

The following factors are considered in the development of a CWBS:

- The CWBS provides a "home" for every direct charge activity on the project and minimizes the difficulty in identifying charges with tasks.
- There is no general rule applied about the relative number of Control Accounts versus contract size. Project organization and the SOW are only two factors considered in development of Control Accounts within the CWBS.
- Each element of the CWBS represents an aggregation of all subordinate elements listed below it.
- All items appearing can be identified to the contract milestone schedule with the exception of some LOE activity.
- Cost can be easily identified to each CWBS element without allocation to two or more elements, and costs are collected and summarized upward without bypassing lower level elements.
- Integration of the lowest levels of the CWBS with the functional organization structure or responsibility structure produces Control Accounts that are not too small (i.e., the gain in control is not offset by the cost of administration).
- The CWBS allows activities that cut across all deliverable portions of the CWBS. These activities are significant enough to warrant being identified as specific elements.
- Contract line items and end items are included if they are consonant with customer WBS guidance.
- The CWBS includes elements intended to be subcontracted, with identification of subcontractors if known.

4.2.3 Contract Work Breakdown Structure Coding

During development of the CWBS, a coding system is developed for use in summarizing data through all CWBS elements. The CWBS code becomes part of the accounting code used to collect actual costs incurred. A full discussion of the accounting structure is detailed in Section 11, Cost Accounting.

4.2.4 Contract Work Breakdown Structure Index and Dictionary

When the CWBS is fully developed, a CWBS Index, an indentured listing of all CWBS elements, is prepared. Figure 4-3 provides a sample CWBS Index.

The CWBS Dictionary (Figure 4-4) correlates the contractual SOW for the project with specific CWBS element references. Functional Managers usually use the dictionary to ensure that each level of the CWBS represents the tasks specified in the contract SOW and to ensure that each Control Account Work Authorization (CAWA) correctly describes and includes the applicable contractual effort. The following items should be included in the narratives within the CWBS Dictionary for the affected CWBS elements:

- Contract Line Item Numbers and a description of applicable end items should be shown in each applicable CWBS element.
- Each CWBS element to be subcontracted should be annotated as such in the narrative, with identification of major subcontracts and subcontractor's names.

Using the CWBS Dictionary, the Project Office ensures that each CWBS level represents a task specified in the contract SOW. This, in turn, verifies that by managing to CWBS elements, the Project Manager is considering the total contract SOW.

4.3 Organizational Breakdown Structure

4.3.1 Organization

The Project Manager selects functional categories with a direct relationship to the project. These functions are grouped into a project organization chart (Figure 4-5) that identifies all levels of management for every organization responsible for performing a portion of the contract work. The project organization is the framework within which the Project Manager assigns responsibilities for contract tasks.

4.3.2 Organization Coding

Once project organization has been determined, the coding necessary to identify each element is determined. The organizational code is used on authorization documents to identify the functional data summation trail.

4.3.3 Project and Functional Organization

Within LGM INTERNATIONAL, the organizational elements associated with the management of a given contract are the functional organization and the project organization.

4.3.3.1 Functional Organization

The functional organization is the organization through which the individual LGM INTERNATIONAL employees report for supervision and management direction for project performance. Functional organizations (e.g., Operations, Product Assurance, Engineering, Construction) perform the project work as defined and delegated by the Project Office.

4.3.3.2 Project Organization

The project organization is structured according to project size and complexity. Under the direction of the Project Manager, the project organization organizes, coordinates, and controls all work effort on a contract, using LGM INTERNATIONAL EVMS and the Project Plan to define and delegate work to the functional organizations.

The project organization is normally initiated in the proposal phase of the contract and is instrumental in preparing the proposal response. The project organization supports proposal negotiations and is formally authorized after the contract is awarded. To the greatest extent possible, the proposal team is the project organization and identifies potential CAMs.

The project organization is typically composed of the Project Office (usually Project Management, Project Finance, and Project Controls), Project Contract Administration, and a representative from each functional organization involved in the project. The project organization chart depicts a typical project organization (Figure 4-5).

To the extent practical, the project organization is the same organization that performs the work. In all instances, the project organization and the functional organization managers have functional authority and managerial involvement with the work. In instances in which some of the Control Account effort is performed by another organization, the project organization is the primary performing organization (i.e., the organization that contributes 80 percent or more of the effort). Efforts of the secondary performing organizations is planned as separate Work Packages. Thus, all effort is divided into separate Work Packages until at least 80 percent of a Control Account is performed by a single organization. The Project Manager and CAM determine the division of effort to preclude low-valued Control Accounts that are not cost-effective to manage and may not provide significantly more management visibility. Efforts performed by organizations outside the responsible organization may exceed 20 percent if creating two Control Accounts is not cost effective.

4.3.3.3 Project Manager

The Project Manager is responsible for the total project. The Project Manager has responsibility for the organization and integration of contract planning, scheduling, budgeting, work authorization, and cost/schedule control. The Project Manager derives authority from the Product Line President through the Business Development/Contracts group via the Contract Work Authorization (CWA).

4.3.3.4 Contract Work Authorization

The CWA is the only vehicle that can open, close, or revise a project. This document authorizes the commencement of work to be performed to accomplish the contract tasks committed to by the product line. Costs cannot be accumulated against a project until a CWA budget has been established and released with the approved CWA. The LGM INTERNATIONAL Product Line President has the authority to initiate a project effort. All programs are assigned to a specific Project Manager. Figure 7-2 is a sample CWA.

Use or disclosure of data contained on this sheet is subject to the restriction on the title page of this document.

Page 27

4.3.3.5 Control Account Manager

The CAM has project authority for the management of all personnel and material resources required to accomplish the Control Account work effort. The CAM is responsible for the schedule and associated budget for all work performed within the CAM's Control Account. The CAM reports through the Functional Manager to the Project Manager in all matters related to the project responsibilities assigned to the CAM. The relationship among the Project Manager, Functional Manager, and the CAM is one of accountability and is established when the Project Manager, CAM, and the CAM's Functional Manager agree on the scope of work, schedule, and budget for a particular project effort.

4.3.3.6 Functional Manager

The Functional Manager usually is responsible for supervision and is involved in the performance appraisal and salary administration of the CAM. The Functional Manager provides resources to enable the CAM to perform the assigned project effort; is directly involved in the negotiation of project effort during the negotiation of the Control Account, and is responsible for reviewing, on a monthly basis, the status of all efforts being performed by the CAM at a summary level. The Functional Manager is kept aware of specific out-of-tolerance Control Account variances, planned corrective actions, and the impact of these variances on a monthly basis. The Functional Manager also is responsible for developing functional VARs when required.

4.4 Responsibility Assignment Matrix

4.4.1 Interrelationship of CWBS and OBS

The CWBS defines and organizes the work to be performed. The organization structure reflects how personnel are organized and who will be responsible for the work. To assign work responsibility to appropriate organizations, a RAM (Figure 4-6) must show functional responsibility for a specified element of work (see Figure 4-6). The purpose of the RAM is to illustrate the intersection points of the CWBS and the OBS where the performance of work is managed. This results in the definition of Control Accounts. The process of defining Control Accounts is the responsibility of the Project Manager and Project Controls, working in conjunction with the functional organization. Each Control Account is given a number to identify the budget for the work planned and to identify and accumulate the ACWP.

The RAM is an essential tool for the definition of Control Accounts at the lowest levels of the CWBS and the organizational structure.

4.4.2 Major Subcontractors

Major subcontracts are identified and discussed in Section 10.

4.5 Development of the CWBS Dictionary and Responsibility Assignment Matrix

Figure 4-1 is a flowchart that illustrates the process of developing the CWBS Index and Dictionary and the RAM. The following narratives accompany the flowchart:

A. Upon receipt of the contract, the Project Manager and Project Controls, with inputs from Functional Managers and CAMs, work together to extend the CWBS to the lowest level for management control purposes.

B. Project Controls prepares a CWBS Index, or indentured list, of all CWBS elements (see Figure 4-3).

C. Using the CWBS Index as a guide, a CWBS Dictionary description is prepared with inputs from Functional Managers and CAMs for each element (see Figure 4-4).

D. The CWBS Index and Dictionary are submitted to the Project Manager for approval. If they are not approved, they are returned to Project Controls for corrections and resubmitted to the Project Manager (return to Step A). When the CWBS Index and Dictionary are approved, they are returned to Project Controls to be used in the preparation of the preliminary RAM.

E. Based on the CWBS Index and Dictionary and the proposed organization structure, a preliminary RAM is developed by Project Controls. The preliminary RAM may only have major functional organizations identified along the

Use or disclosure of data contained on this sheet is
subject to the restriction on the title page of this document.

Page 28

vertical axis at this point. The CWBS Index and Dictionary and the preliminary RAM are forwarded to the appropriate level of functional management.

F. Functional Managers review the CWBS Index and Dictionary and preliminary RAM. Functional managers, with assistance from Project Controls, is responsible for extending the organizational structure axis of the RAM to the appropriate level by identifying the functional level at which work will be performed. This level represents the organization responsible for the Control Account effort. Functional Managers also identify the CAM responsible for specific Control Accounts.

G. Functional Managers submit the extended RAM information to Project Controls for review. If concerns are identified, RAM inputs are returned to the functional organizations for iteration.

H. Project Controls prepares the final proposed RAM and submits it to the Project Manager for review and approval.

I. When the RAM is approved, Project Controls enters the RAM into the database and distributes copies to Functional Managers.

4.6   Management System Integration

The integration of management control systems, from the contract level through the CWBS and organization levels to the Work Package level, is illustrated in Figure 4-7.

4.7   Control Account Management

4.7.1   Control Account

The Control Account, the natural control point for cost/schedule planning and control, is formally authorized via a CAWA (Figure 7-4). The Control Account scope of work directly relates to a single element of the CWBS. The Control Account represents the level where budgets are authorized for commitment and expenditure of direct labor, material, and ODCs. All work performed on a contract is authorized on a CAWA and assigned to a single functional employee, a CAM, who is accountable for completing the scope of work identified. The SOW that appears on the CAWA is written by the CAM and based on the CWBS Dictionary narrative for the referenced CWBS element. The scope of work, schedule, and budget for the Control Account is agreed on by the Project Manager, the CAM, and the Functional Manager prior to issuance of the final CAWA.

The Control Account is the vehicle for planning, scheduling, budgeting, work authorization, and cost accumulation in a project. Budgeted Cost for Work Scheduled (BCWS), Budgeted Cost for Work Performed (BCWP), and ACWP are available for performance measurement and analysis at the Control Account level. Much of this information is developed/displayed on the Control Account Plan (CAP), an integral part of the CAWA process. The Control Account, therefore, is the point at which the Project Budget Planning, Scheduling, Work Authorization, Performance Measurement, Control Accounting, Data Analysis, and Revision subsystems integrate with each other, the CWBS, and the organizational structure. The establishment of Control Accounts should follow these guidelines (any exceptions to the following require Project Manager approval):

☐   The responsible organization is involved in the management of the resources to perform the work and is held accountable for performing the work.

☐   The effort may not be allocated between two or more Control Accounts. When the effort involved is substantially common to two or more CWBS elements, the Control Account should be assigned to one of the elements.

☐   LOE and discrete work should normally be segregated by Control Account. However, in some cases, LOE and discrete work may be intermingled within the same Control Account. To preclude distortion of performance measurement data, this should be done so that the total amount of one type of effort does not exceed 20% (i.e., there can be no more than 20% LOE in a discrete Control Account and vice versa). In some cases, small Control Accounts may exceed this guideline. When a Control Account exceeds this guideline, the fact that this guideline is exceeded must be acknowledged within the body of the CAWA and noted that it is overridden because of the size of the Control Account.

Use or disclosure of data contained on this sheet is subject to the restriction on the title page of this document.

Page 29

4.7.2   Cost Collection

The Control Account is the lowest required level for the collection of actual costs; however, LGM INTERNATIONAL EVMS may collect costs at the Work Package level. Traceability, in any case, is maintained.

4.7.3   Types of Effort

Because work accomplishment is measured in the same manner in which it was planned, work is planned and defined as one of three basic types: measurable effort, LOE, or apportioned effort.

4.7.3.1   Measurable Effort

Measurable effort is effort in which the total task to be performed has a definite end product and can be subdivided into smaller Work Packages with outputs upon which a measure of work accomplishment can be based.

4.7.3.2   Level of Effort

Efforts designated as LOE contain no clearly definable end product and are generally support or management in nature.

4.7.3.3   Apportioned Effort

Apportioned effort contains "factored work" that is clearly and historically related to measurable effort, such as the relationship of quality engineering to production engineering.

4.7.4   Work Package

4.7.4.1   Work Package Development

Work Packages are the subdivision of work effort below the Control Account level. The task must be clearly defined, scheduled, budgeted, and assigned to a single performing organization responsible for its completion. Performance measurement techniques are assigned to each Work Package, and the package is identified by its cost elements (e.g., labor, material, ODCs).

4.7.4.2   Duration

The duration of Work Packages should be short to reduce the problems associated with determining the value of completed work. Work Packages should be no longer than 2 to 3 months in duration. Larger Work Packages should contain intermediate milestones to prevent distortion of performance measurement, which is described in more detail in Section 8.

4.7.4.3   Types of Work Packages

☐ Discrete, which produce a specific end product or end result and tangible and measurable output (e.g., fabrication of one or more parts)

☐ Apportioned, which involve factored effort directly related to other discrete tasks (e.g., inspecting, drafting checks)

☐ LOE, which do not produce an end product (e.g., project management, surveillance, liaison)

All contract work must eventually be planned as one of these categories during contract performance.

LOE and Apportioned Work Packages should be kept to less than 20% of any measurable Control Account to avoid distorting performance measurement of discrete tasks. It is preferable, where possible, to make all Work Packages within a Control Account the same type. When the value of LOE and/or Apportioned Work Packages cannot be kept below 20%, the Control Account may be broken into two separate Control Accounts. This division is the decision of the Project Manager, Functional Manager, and CAM, precluding low-valued Control Accounts that are not cost-effective to manage and may not provide significantly more management visibility.

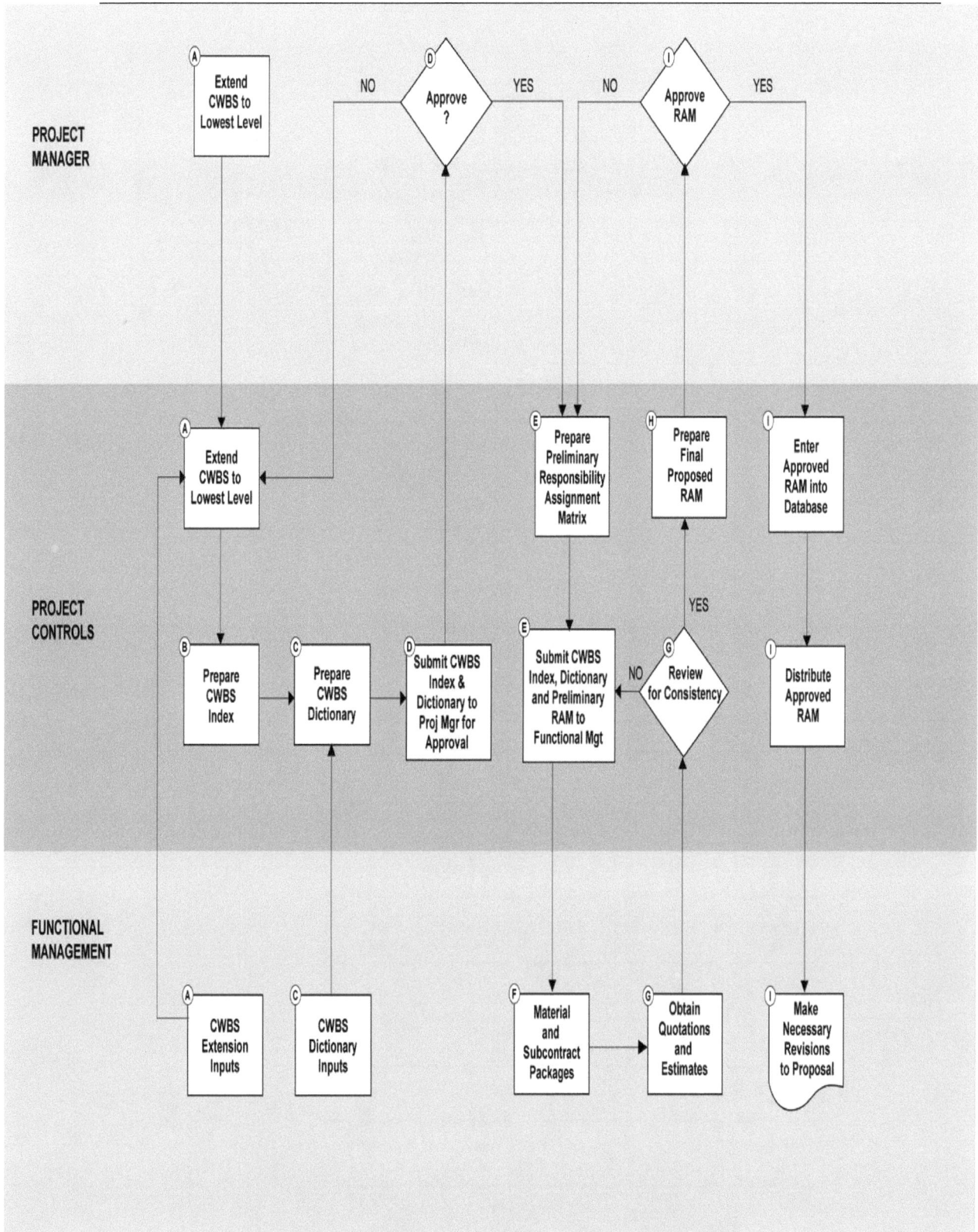

Figure 4-1   CWBS and RAM Development Flowchart

se or disclosure of data contained on this sheet is
subject to the restriction on the title page of this document.

Page 31

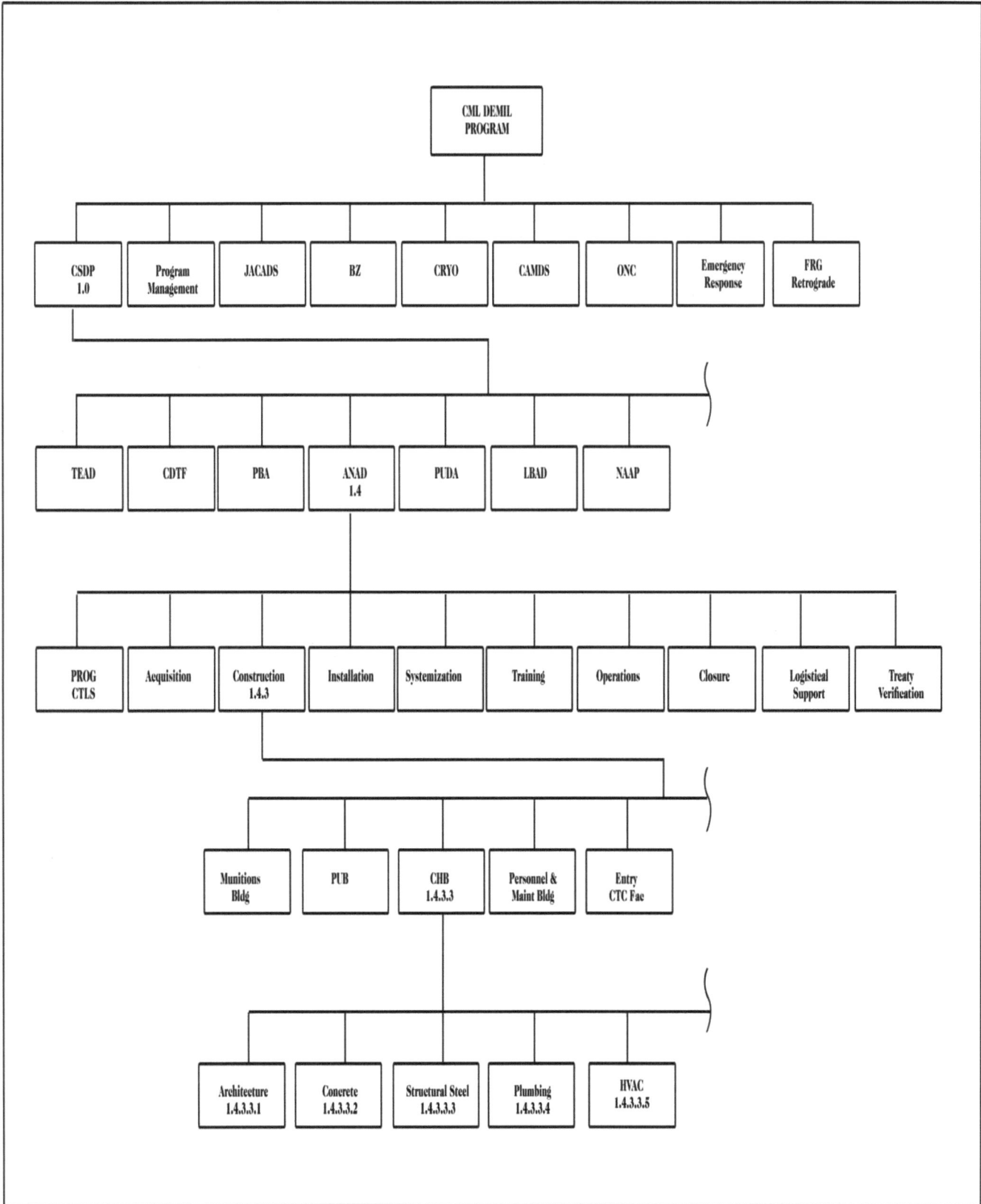

Figure 4-2   Sample Contract Work Breakdown Structure

LINE NO.	LEVEL						WBS ELEMENT	WBS CODE	CLIN	SOW PARA	SPECIFICATION
	1	2	3	4	5	6					
1	X						ANAD	1.4	ALL	ALL	327/875/576
2		X					PROGRAM CONTROL	1.4.1			
26		X					CONSTRUCTION	1.4.3	0003		875
27			X				MINUTIONS BUILDING	1.4.3.1	0003AA	3.0	875
28				X			ARCHITECTURAL	1.4.3.1.1	0003AA	3.0	875
29				X			CONCRETE	1.4.3.1.2	0003AA	3.0	875
30				X			STRUCTURAL STEEL	1.4.3.1.3	0003AA	3.0	875
31				X			PLUMBING	1.4.3.1.4	0003AA	3.0	875
32				X			HEATING, VENT. & A/C	1.4.3.1.5	0003AA	3.0	875
33				X			FIRE PROTECTION	1.4.3.1.6	0003AA	3.0	875

Figure 4-3   Sample CWBS Index

CONTRACT WORK BREAKDOWN STRUCTURE		
ITEM NO	PART II – WBS DICTIONARY AND CONTRACT REQUIREMENTS	

WBS ELEMENT WBS TITLE 1.1.6.0.1 System Project Management	SPECIFICATION NUMBER N/A	SPECIFICATION TITLE N/A
SOW NUMBER 3.1.1.1	CONTRACT LINE ITEM All	CONTRACT END ITEM/ QTY DATE ITEM
DATE: 19 NOV 03 REVISION NUMBER: 1	COST CONTENT: WBS CODE 1.1.6.0.1	WORK ORDER/ WORK AUTHORIZATION Reference Appendix _____
REVISION AUTHORIZATION:	COST DESCRIPTION	
APPROVED CHANGES:	SYSTEM CONTRACTOR	
ELEMENT TASK DESCRIPTION TECHNICAL CONTENT All management effort associated with systems engineering and technical control as well as the business management of the PROJECT XYZ.	1. Encompasses the planning, directing, controlling the definition, development and production of the system including the functions of logistics and logistics ground support, facilities, personnel training, testing and acceptance of the support equipment. 2. Perform cost/schedule control management, contract administration and subcontractor management to accomplish overall project objectives. 3. Perform all efforts associated with facilities planning, project reviews, maintenance of a technical library on specifications and coordination of interfaces to meet contractual requirements.	
APPLICATION WORK STATEMENT NARRATIVE	ASSOCIATE/SUB/SUBCONTRACTOR	

Figure 4-4   Sample CWBS Dictionary

se or disclosure of data contained on this sheet is
subject to the restriction on the title page of this document.

Page 34

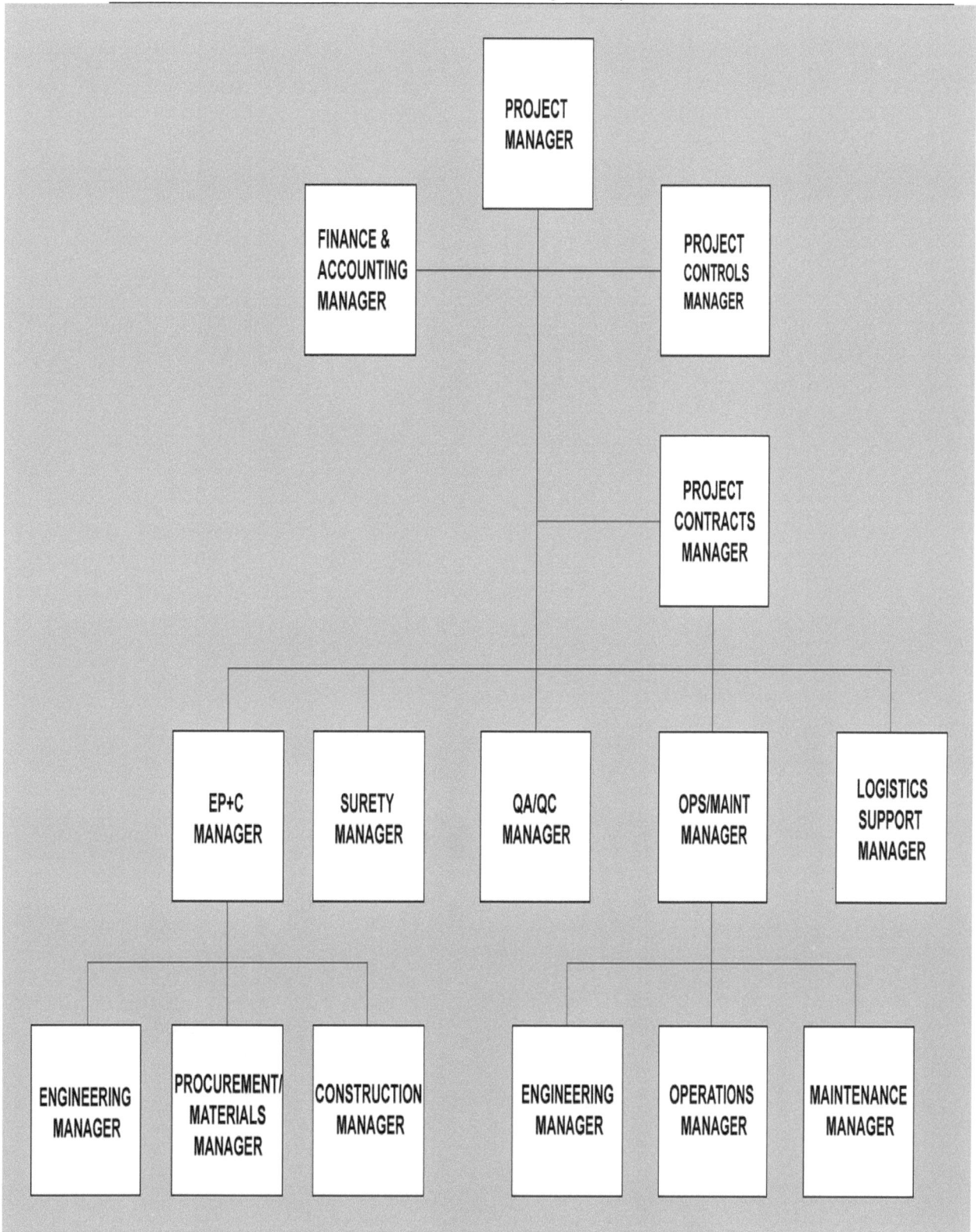

Figure 4-5   Sample Project Organization

se or disclosure of data contained on this sheet is
subject to the restriction on the title page of this document.

Page 35

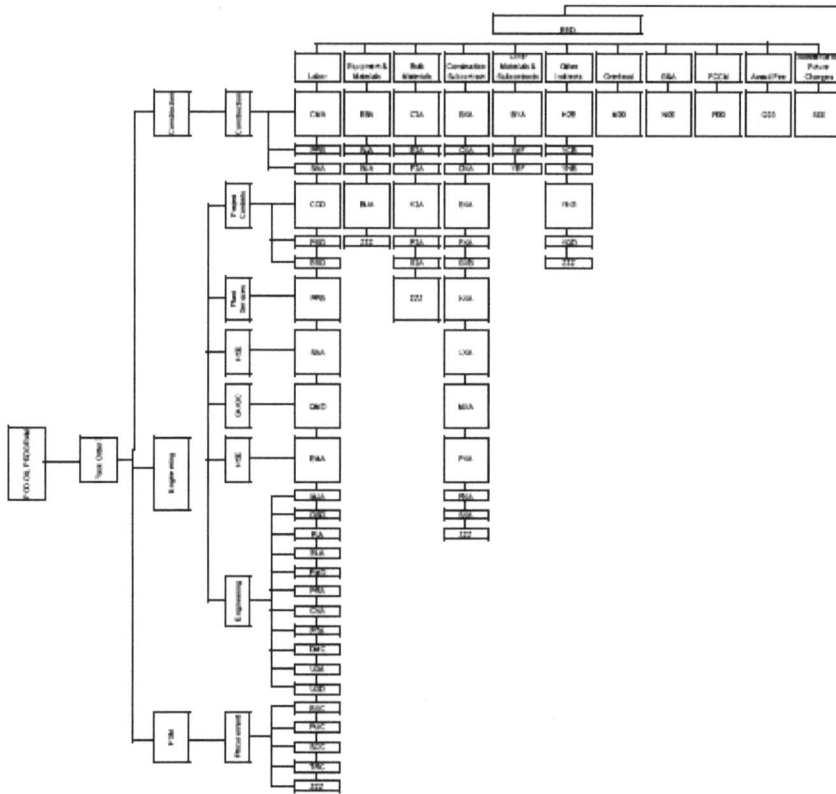

Figure 4-6   Sample Responsibility Assignment Matrix

CWBS LEVEL	ORGANIZATION LEVEL	SCHEDULING	BUDGETING	WORK AUTHORIZATION	PERFORMANCE MEASURE
Contract	Project Manager	• Contract Work Authorization (CWA) • Project Master Schedule	Contract Work Authorization (CWA)	Contract Work Authorization (CWA)	Cost Performance Report (CPR)
Reporting Level	Functional Manager	Intermediate Schedule	Project Work Authorization (PWA)	Project Work Authorization (PWA)	• Cost Performance Report (CPR) • Internal Monthly Performance Reports Variance Analysis Reports (VAR)
Control Account	Control Account Manager	Detail Schedules Control Account Plan	• Control Account Work Authorization (CAWA) • Control Account Plan	Control Account Work Authorization (CAWA)	Internal Monthly Performance Reports Variance Analysis Reports (VAR)
Work Package	Control Account Manager	Control Account Plan	Control Account Plan	Control Account Plan	Internal Monthly Performance Reports

Figure 4-7   Management System Integration

5. Project Budget Planning

Use or disclosure of data contained on this sheet is
subject to the restriction on the title page of this document.

Page 37

5.1 Overview

The process of project budget planning results in time-phased budget baselines against which monthly cost and schedule performance are measured and establishes the relationships of the contract scope of work to the budget and schedule.

5.2 Proposal/Budget Relationships

The funding and budget available to a project are determined in large part by estimates made during proposal preparation and modified during contract negotiation. A budget estimate (proposal) and a firm budget (contract) are prepared in basically the same manner. The amount and classification of labor required to complete each Work Package is determined and defined in terms of labor hours, which are converted to labor dollars and burden dollars. Costs of purchased materials and subcontracts, as well as costs of services and computer time, are added. After contract negotiation and all contract work is definitized, the Contract Target Cost (CTC) becomes the Contract Budget Baseline (CBB).

5.3 Establishing the Contract Baseline

In establishing a contract scope of work, schedule, and budget baseline, five elements are defined.

☐ A clearly defined contractual scope of work

- Describes the authorized work and any special contractual provisions

☐ An integrated and traceable contract budget

- Establishes budgets for all authorized work with separate identification of cost elements (e.g., labor, material, ODCs, overhead)

- Establishes budgets, to the extent possible, that can be identified in discrete, short-span Work Packages and establishes budgets for this work in terms of dollars, hours, and/or other measurable units

- Identifies far-term effort in larger planning packages for budget purposes when the entire Control Account cannot be subdivided into detail Work Packages

- Identifies MR and Undistributed Budget (UB)

☐ An integrated and traceable contract schedule

- Schedules the authorized work in a manner that describes the sequence of work and identifies the significant task interdependencies required to meet the development, construction, operations and maintenance, and/or delivery requirements of the contract

- Establishes and maintains a time-phased budget baseline at the Control Account level against which contract performance can be measured

- Establishes, to the extent possible, work that can be identified in discrete, short-span Work Packages, and establishes budgets for this work in terms of dollars, hours, and/or other measurable units

☐ CAPs that integrate the other elements of the baseline

- Identify physical products, milestones, technical performance goals, and other indicators that will be used to measure output

- Ensure that the sum of all Work Package budgets plus Planning Package budgets within a Control Account equals the Control Account budget

- Ensure that the CTC plus the estimated cost of Authorized Un-Priced Work (AUW) is reconciled with the CBB

- A budget structure that has the concurrence of the responsible organization as described in Section 7.

Use or disclosure of data contained on this sheet is subject to the restriction on the title page of this document.

Page 38

5.4   Budget Baseline Management

The Project Office maintains a CBB log is maintained for each contract. The CBB log is used to display at any point in time the overall budget status of the project (i.e., the UB, the Performance Measurement Baseline [PMB], the MR). The CBB log is the vehicle by which project budget releases will be tracked and displayed. The CBB log displays MR (a reserve not allocated for any known tasks, which can be used by the Project Manager to budget activities that arise during the duration of the project within the scope of the contract), UB (budget allocated for known tasks not yet released to specific Control Accounts), and Control Account budgets.

All contract changes affecting the CBB are tracked, as well as the planning and budgeting that follows such changes. Internal replanning (to/from Control Accounts, UB, or MR) resulting from contract changes can be tracked to the associated change within the CBB log.

The time-phased PMB values remain unchanged except for the following:

- Contract changes
- Transfer of budget from MR
- Assignment of UB to applicable Control Accounts
- Revision request process described in Section 14 consisting of
    - Replanning
    - Budget transfers – Budget Transfers requires the scope to be transferred with the budget.
- Reprogramming

5.5   Budget Definitions

The budgeting process yields a hierarchy of budgets and budget responsibilities. This hierarchy is illustrated in Figure 5-1. Each block in the hierarchy is briefly described below.

A. Authorized Un-Priced Work
    Authorized work from the customer that has not been negotiated and definitized, but for which authority to proceed with planning and accomplishing the work has been received. The budgeted value of authorized work may be based on the contractor's proposal for new work.

B. Contract Target Price
    This value is the total negotiated contract value. It includes the negotiated CTC plus planned profit or fee.

C. Contract Target Cost
    This value is the total negotiated contract cost. It excludes planned profit or fee.

D. Contract Budget Base
    Contract Target Price less fee or profit yields the CBB. This is the starting point in the internal budgeting process. At the beginning of the contract, the CBB represents the budget authorization of the contract and is based on the negotiated and definitized contract cost (i.e., price less fee). The CBB is always equal to the negotiated cost for definitized work and the estimated cost of all AUW. The CBB is maintained by the Project Office and revised only by customer-directed and/or authorized contract changes. The CBB is always reconcilable to the CTC.

E. Management Reserve
    MR is an amount of contract budget set aside by the Project Manager at the start of the project. It is withheld for project management control purposes rather than designated for the accomplishment of a specific task or set of tasks. MR is not part of the PMB.

F. Performance Measurement Baseline (PMB)
    The PMB is the sum of all allocated budgets against which contact performance is measured. The PMB is the sum of both the Distributed Budget (DB) and the UB. The PMB is equal to the CBB less the MR.

G. Undistributed Budget

The UB is the budget identified to specific contract effort that has not been identified yet to CWBS elements at or below the level specified for reporting to the customer (Control Accounts).

H. Distributed Budget

The DB may comprise overhead budgets and Control Account budgets

1) Overhead Budgets

Overhead budgets, including G&A, are budgeted at the Control Account level for tracking purposes (if directed by the customer) or preferably at the total contract level. In either case, Project Controls (not the CAM) is responsible for analysis of overhead at the total contract level.

2) Control Account Budgets

Control Account budgets have a specified scope of work, a Detailed Schedule, and a time-phased budget. The sum of the time-phased budget represents the total Control Account budget.

I. Cost of Money Budgets

Cost of Money budgets are established, if authorized, in the contract in accordance with the contract, normally at the total contract level.

J. Work Package/Planning Package Budgets

Control Account budgets comprising budgets identified to specific Work Packages and Planning Packages.

K. Total Allocated Budget

The sum of all budgets allocated to the contract. The Total Allocated Budget (TAB) is the sum of the PMB and the MR. The TAB reconciles to the CBB with any quantity and cause differences to be documented.

5.6   Control Account Development

The Project Manager, Functional Manager, and CAM agree on the scope of work to be accomplished and the budget required to support the required effort. The entire contract is planned in time-phased Control Accounts to the greatest extent possible. A RAM (Figure 5-2) indicates the major activities required for Control Account development, the organizational manager responsible, and interfaces or joint responsibilities.

The Project Office, upon receipt of a CWA, logs the authorization and amount. The Project Manager evaluates the CBB and establishes an MR budget. The Project Office places the remaining contract amount into the PMB as UB. Preliminary scopes of work, with schedules, are distributed to the appropriate Functional Managers incorporated on the preliminary Project Work Authorization (PWA) form. The Functional Managers segregate and distribute the preliminary scopes of work, with schedules, to the appropriate CAM for initial planning incorporated on the preliminary CAWA form.

Project Controls and the CAM establish an estimate and time-sequenced logic diagram of the defined SOW. This SOW is documented on the CAWA and is based on the CWBS Dictionary for the respective CWBS element. The CAM is responsible for understanding, interpreting, and helping quantify the SOW and for developing the Work Packages and Planning Packages, budget estimate, and time-phased budget schedule for the Control Accounts in conjunction with Project Controls. The CAM and Project Controls reviews, change as necessary, and agree on the CAP. If the SOW, budget, and/or schedule cannot be agreed on, Project Controls and the CAM will resolve the differences with either the Functional Manager and/or the Project Manager.

The Project Manager and respective Functional Manager review, change as needed, and agree on appropriate budgets and schedule. The Project Manager reevaluates the MR amount. Upon final discretion of the Project Manager, the budgets are transferred from the UB to the DB by Control Account. The approved budget, scope, and schedule are distributed to the Functional Managers on the PWA form as depicted in Figure 7-3. The approved budget and scope are distributed to the CAMs on the CAWA form as depicted in Figure 7-4.

If, during any review cycle, changes are made, the cycle will restart at the CAM level. A more detailed description and flow are described in Section 7.

5.7 Control Account Planning

The assigned CAM will accomplish the following three parallel planning tasks:

☐ Prepare a plan for the work to be accomplished based on the SOW provided by the Project Manager. The work should be planned in as discrete a manner as possible. This is accomplished by quantifying as much work as possible into labor hours, quantities, time, etc. Discrete Work Packages are planned identifying milestones, tasks, labor hours, item counts, etc. These Work Packages are recorded in the Progress Measurement subsystem or, as appropriate, on a separate document is attached to the CAP.

☐ Prepare a labor sequence of events to ensure schedule integrity to the intermediate level schedules.

☐ Prepare budget estimates in conjunction with Control Accounts if possible to the Work Package and Planning Package level.

5.8 Management Reserve

MR is not a contingency budget. It is not used to absorb the cost of contract changes, past overruns of Control Accounts, or AUW. MR is used for budgets for unplanned tasks within the contract scope of work but out of scope for the Control Accounts.

5.9 Work Package Budgets

Work Packages are a normal subdivision of a Control Account and constitute the basic building blocks used in planning, measuring accomplishment, and controlling direction of the work. A Work Package is simply a lower level task or work assignment. It represents units of work at the level where work is performed, has a scheduled start and completion date (with milestones if applicable) based on physical accomplishment, and a budget in terms of dollars. The dollar budget may be established in quantities (labor hours, number of material items, or time durations) and converted to dollars in the Control Account budget. When learning curves are used, budgets are time phased, reflecting this learning. The Work Package task description is sufficiently detailed to constrain the application of budget for future effort to the Work Package. A Work Package is characterized by a duration that is limited to a relatively short, manageable, realistic span of time or divided into milestones to facilitate objective measurement; is integrated with overall project Control Account schedules; and describes discrete tasks that define the Work Package.

5.10 Planning Packages

Planning Packages describe work within a Control Account that will occur in future time periods which are beyond those established Work Packages. There may be one, several or no Planning Packages in a Control Account. Planning Packages have a scope of work and a time-phased budget expressed in terms of labor hours, labor dollars, material dollars, etc. Planning Packages must be defined by one or more tasks, each with an associated schedule and/or budget. Planning Packages are normally much larger than Work Packages, but still must relate to specific tasks. Specific Planning Package tasks do not have to be stated in as much detail as is required for tasks in Work Packages. When a Planning Package task is converted into a Work Package, the scope and associated budget is converted to the extent practicable.

5.11 Rolling Wave Planning

Rolling wave planning, a continuous process throughout the contract period, represents the progressive subdivision of gross or aggregate contract planning into component parts (Work Packages). Initially gross or aggregate contract planning is defined as the UB. Subsequently, the process develops into the establishment of the DB. The DB amount is the first refinement to the contract plan. The UB amount is yet to be refined in the contract plan. Rolling wave planning ensures that sufficient budget is available over the contract life span and that front loading is minimized or eliminated. For the contract, budgets are established in time-phased Control Accounts to the greatest extent possible. Rolling wave planning is depicted in Figure 5-3.

Use or disclosure of data contained on this sheet is subject to the restriction on the title page of this document.

Page 41

Rolling wave planning relates to the time-phased conversion of Control Account planning packages to Work Packages. Prior to the beginning of any effort within a Control Account, Work Packages are defined in detail at least 6 accounting months into the future. The effort beyond the 6-month window within a Control Account may be either detailed Work Packages or Planning Packages. In all cases, Work Packages are planned as far in advance as possible. Each Work Package is planned to its logical conclusion.

The Project Manager may override rolling wave planning in the future if a significant contact milestone makes it impractical to plan beyond the normal rolling wave planning window. In these cases, such as at a significant planned customer review or decision point, the Project Manager normally issues guidance in the form of a project directive to accomplish only detailed Work Package planning up to the scheduled decision point.

Use or disclosure of data contained on this sheet is
subject to the restriction on the title page of this document.

Page 42

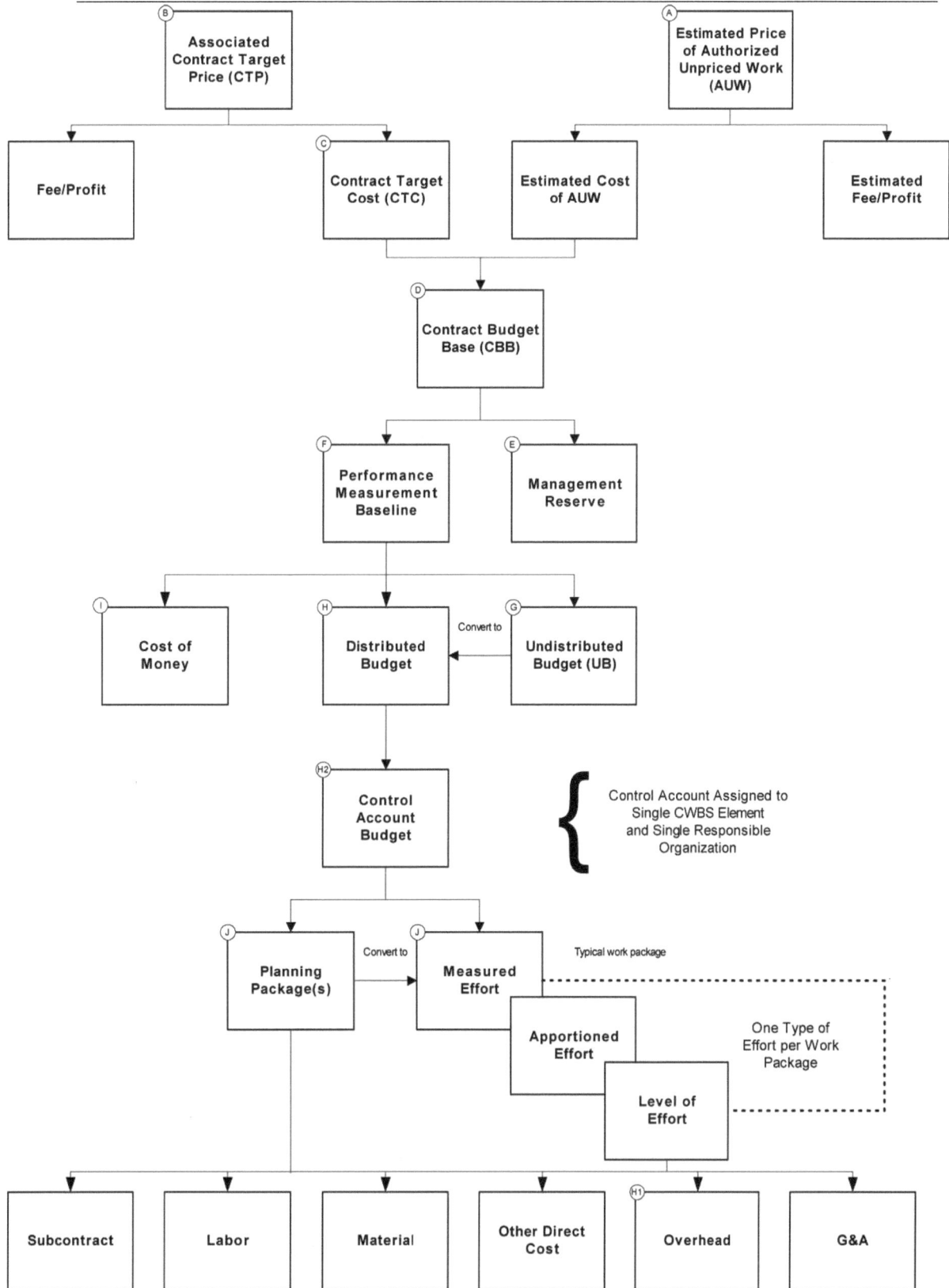

Figure 5-1   Budget Distribution Flowdown

CONTRACT WORK BREAKDOWN STRUCTURE / ORG	PROJECT MGMT 1.0				CONCEPT ENG'G 2.0			PRODUCTION ENG'G 3.0					
	1.1 Proj Mgmt	1.2 Prom Control	1.3 Proc/Matte l Mgmt	1.4 Document Control	2.1 Process Eng'g	2.2 Spec Elec Studies	2.3 Economic Eval Studies	3.1 Site	3.2 Major Equipment	3.3 Piping	3.4 Electrical	3.5 Control Bldg	3.6 Instrument
PROJECT OFFICE													
100 PROJECT MANAGER	X												
172 PROC/MAT'L MGMT			X										
146 PROJ CONTROLS		X											
168 DOC CONTROL				X									
ENGINEER DEPT													
020 PROCESS ENGINEERING					X		X						
040 ELECTRICAL						X					X		
030 ARCHITECTURAL												X	
061 CIVIL								X					
050 MECHANICAL									X				
081 PIPING										X			
070 INSTRUMENTATION													X

Figure 5-2   Sample Responsibility Assignment Matrix

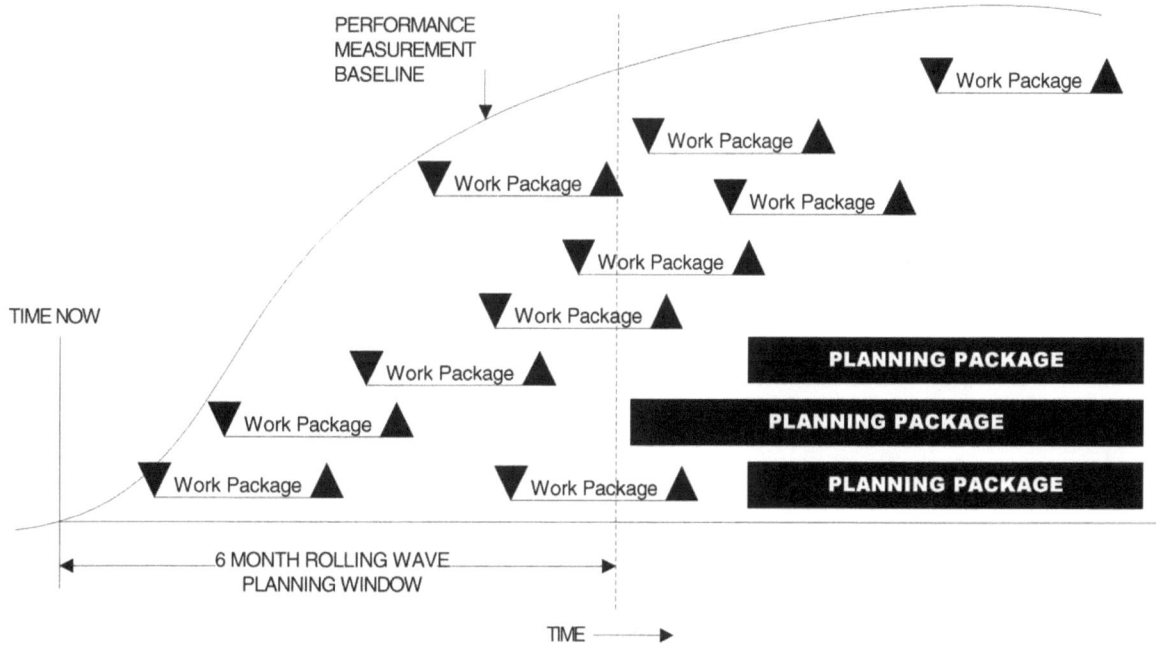

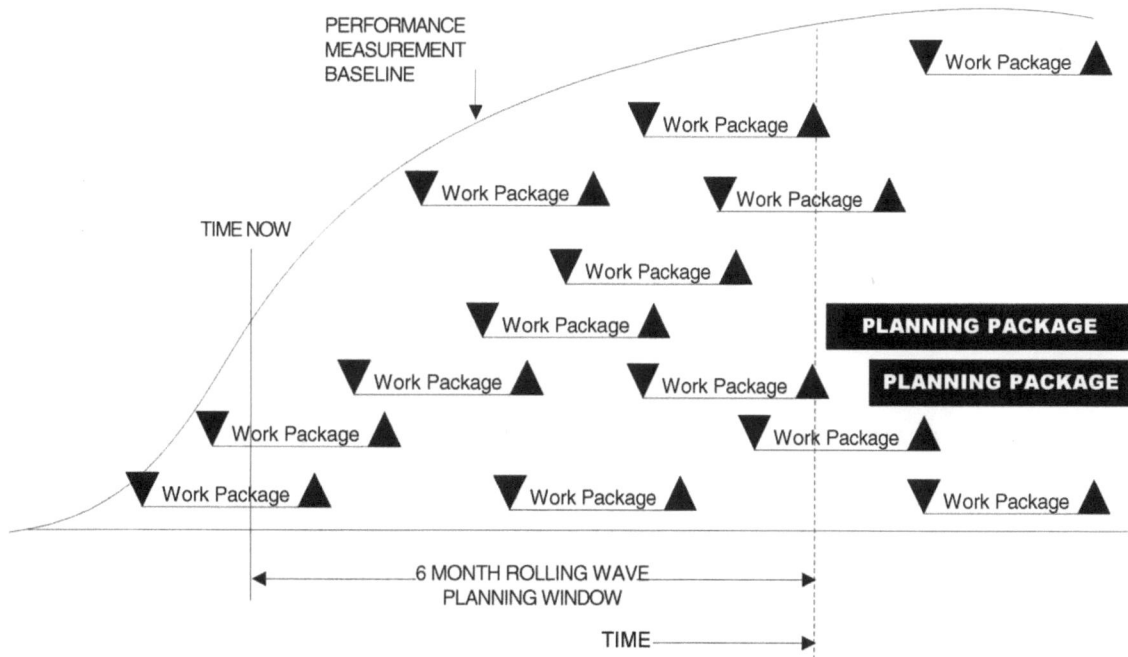

Figure 5-3   Rolling Wave Planning

Use or disclosure of data contained on this sheet is
subject to the restriction on the title page of this document.

Page 45

6. Scheduling

Use or disclosure of data contained on this sheet is
subject to the restriction on the title page of this document.

Page 46

6.1   Overview

The LGM INTERNATIONAL EVMS is a scheduling subsystem that provides a disciplined framework for the development and maintenance of contract baseline schedules that are used to provide time-phased schedule direction from the Project Office to functional organizations responsible for performing the contract work. It is also used to record and report both actual and forecast completions of scheduled milestones and activities. Using its inherent features for vertical and horizontal traceability, LGM INTERNATIONAL EVMS furnishes the ability to identify critical milestones and supporting activities, summarize Detailed Schedule data to summary project levels, and evaluate the impact of current schedule performance on future activities and events. The scheduling subsystem consists of a tiered hierarchy of schedules that start with the top- level PMS and progress in increasing detail to schedules defining Work Package effort assigned to the project organization.

Schedules are developed and maintained for all LGM INTERNATIONAL contracts. Schedules are prepared to the level of detail necessary to provide time allocation and task phasing needed to coordinate the work to meet contractual obligations. The number of schedules and the degree of schedule detail required to achieve contract schedule control is dependent on the contract's type, size, scope of work, and complexity.

The scheduling subsystem is initiated upon receipt of a RFP and is used to define the time frame for proposed contractual activities. The schedules developed during the proposal phase are the basis for the proposed time-phased cost estimates. Following contract award, proposed schedules are amended to reflect the negotiated SOW, Contract Work Breakdown Structure (CWBS), contractual delivery requirements, and customer-specified contractual milestones and to incorporate key supporting milestones and activities designated by the LGM INTERNATIONAL Project Manager. Upon final acceptance and approval as baseline schedules, the schedules come under rigid change control wherein all revisions must be approved by the Project Office prior to incorporation. Section 14 describes this change control process in more detail.

6.2   Schedule Hierarchy

The LGM INTERNATIONAL EVMS consists of baseline schedules and special-purpose schedules. Baseline schedules are approved by the Project Office and constitute the official schedule plan against which schedule performance is measured and reported to LGM INTERNATIONAL and the customer. Baseline schedules, once approved by the Project Office, are maintained under strict control of the Project Office and may not be revised without prior authorization by the Project Office. A three-level hierarchy consisting of the PMS, Intermediate Schedules, and Detailed Schedules comprise the baseline schedules for the contract. This hierarchy is illustrated by Figure 6-1.

Special schedules are developed as required based on constraints and time phasing within the baseline schedules. These schedules are used for such purposes as management presentations, where an extreme detailed level below the Work Package level (e.g., hour-by-hour plant shutdown/turnarounds) may be needed.

6.2.1   Baseline Schedules

The PMS is the top-level schedule in the LGM INTERNATIONAL EVMS three-level baseline schedule approach. All other baseline schedules are subordinate to and must support the time-phased activities and milestones depicted by the PMS. Intermediate Schedules are prepared to expand the degree of detail provided by the PMS. Detailed Schedules are an expansion of Intermediate Schedules containing Work Package details directly related to the activities and milestones in the Intermediate Schedules. This disciplined top-down scheduling method enables the capability to trace and evaluate detailed Work Package schedules to higher level schedules and the summation of accomplishment status at the Work Package level through the Intermediate Schedules to the PMS. The unique aspects of each of the three schedule tiers are described below.

6.2.1.1   Project Master Schedule

The PMS is an overall project management summary schedule depicting major project milestones such as contract dates, client and system completion requirements, and key internal milestones and/or activities identified by the Project Manager as warranting tracking at this level. This information is presented in a summary bar chart and/or a summary CPM network. Figure 6-2 is an example of a PMS.

Use or disclosure of data contained on this sheet is
subject to the restriction on the title page of this document.

Page 47

6.2.1.2   Intermediate Schedule

The Intermediate Schedule shows all major activities (i.e., design, procurement, construction drawings, fabrication, installation, operations). It consists of a time-scaled bar chart or a summary CPM network depicting logical relationships among summary activities. The purpose of this schedule is to allow monitoring of project areas and organization progress toward the objectives for contract compliance. An example of an Intermediate Schedule is shown in Figure 6-3.

6.2.1.3   Detailed Schedule

The Detailed Schedule is a schedule of Work Packages within the Control Account. This schedule is detailed during the Control Account planning activity by the CAM and represents the baseline schedule at a detailed level. The Work Package timelines in the Detailed Schedule are vertically traceable to the Intermediate Schedule. Work Package detail may be represented in the CPM by more summary activities. In any event, the baseline schedule for the Control Account Work Package is represented in the Detailed Schedule. If detailed Work Package schedules from the CAP are carried in the CPM itself, the Detailed Schedule illustrates this schedule detail (Figure 6-4). Figure 6-5 illustrates the vertical traceability of the schedule levels.

6.2.2   Special Schedules

Throughout the performance of the contract there are recurring requirements for the preparation of schedules to provide schedule visibility for special purposes (e.g., management briefings, functional organization controls, corrective action plans, shop loading, facility or equipment utilization, detailed operating plans). Similarly, there is the need for aggregation of schedule data related to a specific subject (e.g., drawing and specification releases, data item submittals, consolidated procurement schedules, project review schedules, significant event coordination schedules). These types of schedules are developed on an as-needed basis for detailed control of critical or priority items. They may contain selected schedule information from official baseline schedules and are updated continually to be consonant with baseline schedule data or may be one-time usage documents depending on their purpose.

6.3   Scheduling Techniques

Various scheduling methods are used to best fit the scheduling requirements of each project. These may include one or more of the following:

☐   Milestone

☐   Gantt

☐   CPM (Arrow Diagram Method - ADM and Precedence Diagram Method - PDM)

☐   Tabular listing

Schedules may be prepared manually or computer generated. Customer-specified scheduling techniques and schedule status reporting formats are used as required, but additional methods may be employed to satisfy LGM INTERNATIONAL EVMS internal requirements. However, all project work at LGM INTERNATIONAL is scheduled in a manner that describes the sequence of work and identifies the interdependencies required to meet contract requirements.

6.4   Schedule Traceability

The Project Office Scheduling Section is responsible for maintaining baseline schedule traceability vertically, horizontally, and historically.

6.4.1   Vertical Traceability

Vertical traceability provides the ability to identify and define the time-phased relationships among scheduled activities and events depicted on CAPs and those at, and through, a higher summary level shown on Intermediate Schedules to the PMS. Figure 6-5 illustrates vertical traceability.

Use or disclosure of data contained on this sheet is
subject to the restriction on the title page of this document.

Page 48

Using the CWBS framework, coupled with the schedule system top-down planning approach, ensures inclusion of vertical traceability during the development of baseline schedules. The Project Office Scheduling Section establishes and maintains vertical traceability among the PMS, Intermediate Schedules, and Detailed Schedules in the baseline schedule hierarchy. This is accomplished through the use of coding structures within the detail activity elements of the CPM schedule, which allows automatic summarization through the CWBS and OBS. In addition, the Project Office verifies the vertical traceability among Control Account schedules and related Intermediate Schedules as part of its review and approval of CAPs prepared by the individual CAMs.

6.4.2 Horizontal Traceability

Horizontal traceability defines time-phased end-to-end relationships among activities and milestones and identifies required precedence activities and constraints that control start and completion of scheduled work. It enables the determination of critical paths and evaluation of the effects of current schedule performance status on activities and milestones scheduled for accomplishment in the future.

The Project Office Scheduling Section is responsible for ensuring that horizontal traceability is incorporated into all contract baseline schedules. Various methods may be used to establish and document horizontal traceability. These techniques may range from simple depiction of constraints on Gantt and milestone schedules to application of extensive logic diagrams requiring support of computer-aided schedule evaluation processes. Selected techniques depend on the size, duration, and complexity of the project. When a CPM schedule is developed at the Work Package level, the CAMs and the Project Office Scheduling Section identify interrelationships and interdependencies at this level and between Control Accounts where appropriate.

6.4.3 Historical Traceability

Historical traceability provides an audit trail for all revisions to baseline schedules. All baseline schedule revisions are documented on an approved revised CAWA and maintained on file in the Project Office. This process is used to identify and record each schedule revision, the date and purpose of its incorporation, and the authority for the revision action.

Revisions to Control Account schedules and related Planning and Work Package schedules are documented on revised CAWAs.

6.5 Proposal Schedule

Upon receipt of an RFP, a preliminary proposed PMS is prepared under the direction of the Proposal Manager. This schedule is developed to delineate the proposed period of performance, indicate significant milestones for key decision points, and identify contractual deliveries and major constraints and interfaces. It also designates the time phasing of the major activities defined by the top levels of the proposed CWBS. The preliminary proposed PMS is reviewed by the proposal team and functional organization representatives to evaluate reasonableness of durations and validity of interdependencies and constraint logic and to reach agreement on the definition of each major milestone. The preliminary proposed PMS is then amended to incorporate the results of this review and issued by the Proposal Manager as the official proposed PMS, which will be used as the basis for development of more Detailed Schedules needed for preparation of time- phased resource requirement estimates for labor, materials, and ODCs.

Additional schedules may be prepared to meet RFP requirements for schedule visibility for special interest areas such as integrated test schedules, facility activation schedules, subcontractor support schedules, and so forth. All schedules prepared for the proposal are retained by the Proposal Manager for use as support data during contract negotiations and as a basis for development of baseline schedules after contract award.

6.6 Schedule Development

The baseline plan (schedule, CWBS, Control Accounts) for the project are developed with the full participation of the project organization. The plan represents the agreed best approach to accomplish the project's time and budget activities. The project organization is responsible for accomplishing its assigned scope of work and monitoring, appraising, and controlling performance in accordance with the plan. Scheduling and schedule control, therefore, are a shared responsibility from project management to the design engineers, technicians, draftspersons, materials, procurement, construction, operations, and maintenance personnel. In this context, Project Controls provides the expertise and tools to coordinate the development of the plan to ensure communication of plan information and document the plan in the proper format.

Typically, Project Controls provides the available project scheduling information to the project organization. The information includes preliminary CWBS planning worksheets, model networks with project-specific milestones, and assumptions or ground rules critical to plan development. The project organization, consisting of the Functional Manager, CAMs, and other project participants provides input to the scheduling process in the form of task definition, interdependencies, durations, key intermediate milestones, lead times, and so on. Project Controls consolidates the input and, with the Project Manager, coordinates resolution of conflicts. They then agree on a schedule that enables the accomplishment of major milestones.

Project personnel estimate work content, labor hours, and staffing. Project Controls consolidates this information to develop personnel plans and evaluates staffing against the schedule and budgets.

In this planning phase, logical work steps are identified and evaluation is made of predecessor/successor relationships of activities, leading to the establishment of a logical sequence. Alternative options are identified, and a basis for selection is determined. Critical resources (i.e., material, equipment, personnel by skills) are highlighted in relationship with time-phase requirements. Estimates of project activities and duration are established. The planning phase produces a logically ordered activity list and display of project activities.

The Project Manager approves the plan, which becomes the baseline for development of the detailed CAP schedules for accomplishing the work. Project Controls assists the CAMs in developing Control Account schedules and verifies their compatibility. The CAM develops the detailed Work Package schedules in the manner in which the work will be accomplished. The CAM identifies physical accomplishments as events or milestones and the scheduled start and completion dates.

The schedule development process flow is depicted in Figure 6-6.

6.7   Schedule Statusing and Forecasting and Reviews

The CAM provides actual progress schedule position against the schedule baseline and projected completion dates. Significant slippages are highlighted to indicate the need for corrective actions and work-around. With the control point at the Control Account, maximum sensitivity to problems is accomplished and correlated cost impacts are identified.

As the project team moves through the execution phase, the CAM monitors, appraises, coordinates, and reports on progress, problems, and corrective actions.

The CAMs monitor and control the day-to-day execution of the work. They are responsible for meeting technical, time, budget, and quality objectives for the Control Accounts. Project Controls aids in the coordination and communication of progress and problem information. Project Controls continuously evaluates the progress and personnel loading against the Detailed Schedule and flags problems that may impact schedule milestones. Project Controls collects the update information, develops the schedule and progress control analysis, and reports the results to the CAM, who is responsible for identifying all problems, defining corrective actions, and providing this information to appropriate levels of managers for decision making.

Schedule control initially consists of collecting status information relative to activity progress, updating the schedule, preparing forecasts, and comparing results with the baseline.

If significant deviations are found, recovery plans are prepared to demonstrate how the project will be brought back into compliance with the baseline schedule with approved corrective actions.

Status information is collected from the CAM, including actual start and/or completion dates for activities with physical progress during the reporting period and estimates of remaining duration of activities that are in progress as of the report cutoff date (data date). Because activities are resource loaded by performing organization resource type, it may be necessary to revise the resource forecast periodically.

Schedule analysis is performed with each schedule update. This analysis includes discussion of critical or near-critical items and recovery plans that have been developed for areas of the schedule that show negative float and discussion of schedule activities that are late to the baseline schedule, whether or not they are critical in terms of float.

Work progress and accomplishment of milestone objectives are reviewed by the Project Office. Ongoing reviews ensure continuous conformance to established technical and project requirements. Bar (Gantt) charts and S curve graphs are used for management reviews. The schedule status process is shown in Figure 6-7.

6.8   Schedule Changes

All PMSs and Intermediate Schedules are formally approved and issued by the Project Office. Once the schedule baseline is established (i.e., the Project Manager has issued the PMS and Intermediate Schedules), changes can be made, as discussed in Section 14, if the following actions are taken:

☐   Justification for the change is provided to the Project Manager. Because all schedules support the next highest level of schedules, any schedule change proliferates through the scheduling baseline.

☐   The changes are reviewed in detail to ensure that all schedules support the contract and that Intermediate Schedules support the PMS.

☐   If contractually required, changes to the PMS are coordinated with the customer and/or approval is received from the customer.

☐   The revised schedules are formally issued by the Project Manager to Functional Managers and CAMs.

☐   A schedule change is documented on a revised CAWA and a revised PWA if necessary.

☐   Affected Control Accounts are revised as necessary.

This element of schedule control provides the necessary audit trail for all revisions to baseline schedules, ensuring historical traceability.

6.9   Schedule Development Process

The schedule development process flow, illustrated in Figure 6-6, is described below. Although scheduling is part of Project Office responsibilities, it is identified separately in the flowchart to better indicate responsibilities.

A.   The Project Manager receives a CWA from Business Development/Contracts. The CWA establishes the overall contract schedule and provides contract references for additional information. The CWA also authorizes the Project Manager to expend LGM INTERNATIONAL resources to accomplish the contract effort.

B.   The Project Office develops the CWBS and CWBS Index and Dictionary.

C.   Project Office Scheduling refines the PMS based on contract negotiations, including requirements, major milestones, and deliverables.

D.   Project Office Scheduling, working with the Functional Managers and available CAMs, develops preliminary Intermediate Schedules and near-term Detailed Schedules. Once the preliminary Intermediate Schedule is completed, it is forwarded to the Project Office for review and approval.

E.   The Project Office reviews the revised PMS and preliminary Intermediate Schedules for consistency and continuity and, if required, returns the schedules unapproved for iteration by Project Office Scheduling, Functional Managers, and CAMs.

F.   Once the PMS and Intermediate Schedule are approved, they are recorded as baseline schedules in the LGM INTERNATIONAL EVMS database and copies are forwarded to the Functional Managers and CAMs.

G.   Upon receipt of the baseline schedule and PWA, the Functional Manager issues the CAWA to the CAM authorizing the development of the CAP, including detailed Work Package and Planning Package schedules.

H. Upon receipt of the approved baseline Intermediate Schedule and CAWA, the CAM develops the detailed CAP and forwards it to the Functional Manager.

I. The Functional Manager reviews the detailed CAP and returns it to the CAM for iteration as required. Upon acceptance, the Functional Manager forwards the CAP in a package that represents the work authorized by the PWA to Project Office Scheduling.

J. Project Office Scheduling reviews the CAP for consistency and continuity with the baseline schedule and forwards the package to the Project Office.

K. The Project Office reviews the CAP representing the work authorized by the PWA and returns the plans to the Functional Manager and CAM, through the Functional Manager, if not approved for iteration.

L. The approved CAP and BCWS are input into the LGM INTERNATIONAL EVMS database.

6.10   Risk Management Overview

LGM INTERNATIONAL's Project Risk Analysis process addresses threat/opportunity identification, analysis, mitigation and management throughout a project's entire lifecycle. From business acquisition through project execution, the risk process incorporates probabilistic analyses to quantify how risk positively or negatively affects a project's primary objectives and benchmarks.

LGM INTERNATIONAL recognizes risk as something that exists in all situations. As a result, a project team needs to be able to:

- ☐ Identify and acknowledge the presence of risk
- ☐ Eliminate threats where possible
- ☐ Co-exist with threats in the most efficient manner when they cannot be eliminated
- ☐ Manage threats/opportunities proactively

The risk process encompasses a number of principles as follows:

- ☐ Defining qualitative vs. quantitative analyses
- ☐ Understanding probabilistic vs. deterministic forecast outcomes
- ☐ Managing risk as a project management core process along the entire project lifecycle
- • Integrating the risk process within an organization's culture as the natural way of conducting business

LGM INTERNATIONAL's Project Risk Analysis process subscribes to the methodology for identifying threats and opportunities, for compiling a risk register, for developing and analyzing cost and schedule risk models as well as for creating risk deliverables to aid in the ongoing management of risk (see Figure 6-8 for the project risk analysis process diagram).

LGM INTERNATIONAL's project schedules contains many activities or tasks that cumulatively represent the scope of work to be completed. These activities are assigned durations and strung together with logic that indicates sequence and priority. This set of tasks must be completed on time and in the correct sequence to finish the project as planned. This is the "deterministic" schedule, which includes logic and durations that determine which activities are critical to the success of the project.

The deterministic schedule illustrates the requirements for completing the project, but it does not indicate the probability of completing the project as planned. A risk analysis applies probability to the deterministic schedule. The Risk Analysis creates a probabilistic schedule.

The risk management program will review quantitative and qualitative requirements for the program.

The Key components of the Risk Management Program are:

Use or disclosure of data contained on this sheet is subject to the restriction on the title page of this document.

Page 52

- ☐ Schedule
- ☐ Risk Assessment
- ☐ Risk Response Planning
- ☐ Risk Monitoring
- ☐ Recovery Plans

Use or disclosure of data contained on this sheet is
subject to the restriction on the title page of this document.

Page 53

		BASELINE SCHEDULES
LEVEL 1	PROJECT MASTER SCHEDULE	□ SCOPES TOTAL CONTACT PERIOD OF PERFORMANCE □ CONTRACT AND MAJOR PROJECT MILESTONE □ ORGANIZED BY TOP LEVEL CWBS ELEMENT ACTIVITIES • PROJECT MANAGER'S SCHEDULE DIRECTION FOR PROJECT PERFORMANCE
LEVEL 2	INTERMEDIATE SCHEDULES	• EXPANDS PROJECT MASTER SCHEDULE ACTIVITY AND MILESTONE DETAIL • CUSTOMER, FUNCTIONAL ORGANIZATION & MAJOR SUBCONTRACT MILESTONES • FUNCTIONAL ORGANIZATION BY MAJOR CWBS ELEMENT • MAJOR SUBCONTRACTS • CONTRACT DELIVERY SCHEDULES • CWBS/ORGANIZATION ACTIVITY INTERFACES/CONSTRAINTS • BASIS FOR FUNCTIONAL ORGANIZATION CONTROL ACCOUNT SCHEDULE • TYPICALLY SHOWS ONE ACTIVITY PER CONTROL ACCOUNT
LEVEL 3	DETAIL SCHEDULES	• SUPPORT TRACEABILITY TO HIGHER LEVEL SCHEDULES • BASIS FOR TIME-PHASED PERFORMANCE MEASUREMENT BASELINE (BCWS) • THESE SCHEDULE DETAILS MAY BE MAINTAINED WITHIN THE CPM, HOWEVER, ARE GENERALLY REPRESENTED AT A MORE SUMMARY LEVEL WITHIN THE CPM

		SPECIAL SCHEDULES

SPECIAL PURPOSE SCHEDULES	SPECIAL PURPOSE SCHEDULES	• SCHEDULES PREPARED FOR A SPECIAL PURPOSE OR A PARTICULAR INTEREST, I.E,
		  – MANAGEMENT PRESENTATIONS RECEIPT SCHEDULES – MATERIAL
		  – CONTINGENCY SCHEDULES – SPECIAL STUDIES
		  – OPERATIONAL RECOVERY PLANS SCHEDULES – SHOP LOAD
		  – DETAIL SUPPORT SCHEDULES OPERATING PLANS – CURRENT
		  – ENGINEERING RELEASE SCHEDULES
		NOT PART OF HIERARCHY OF BASELINE SCHEDULES

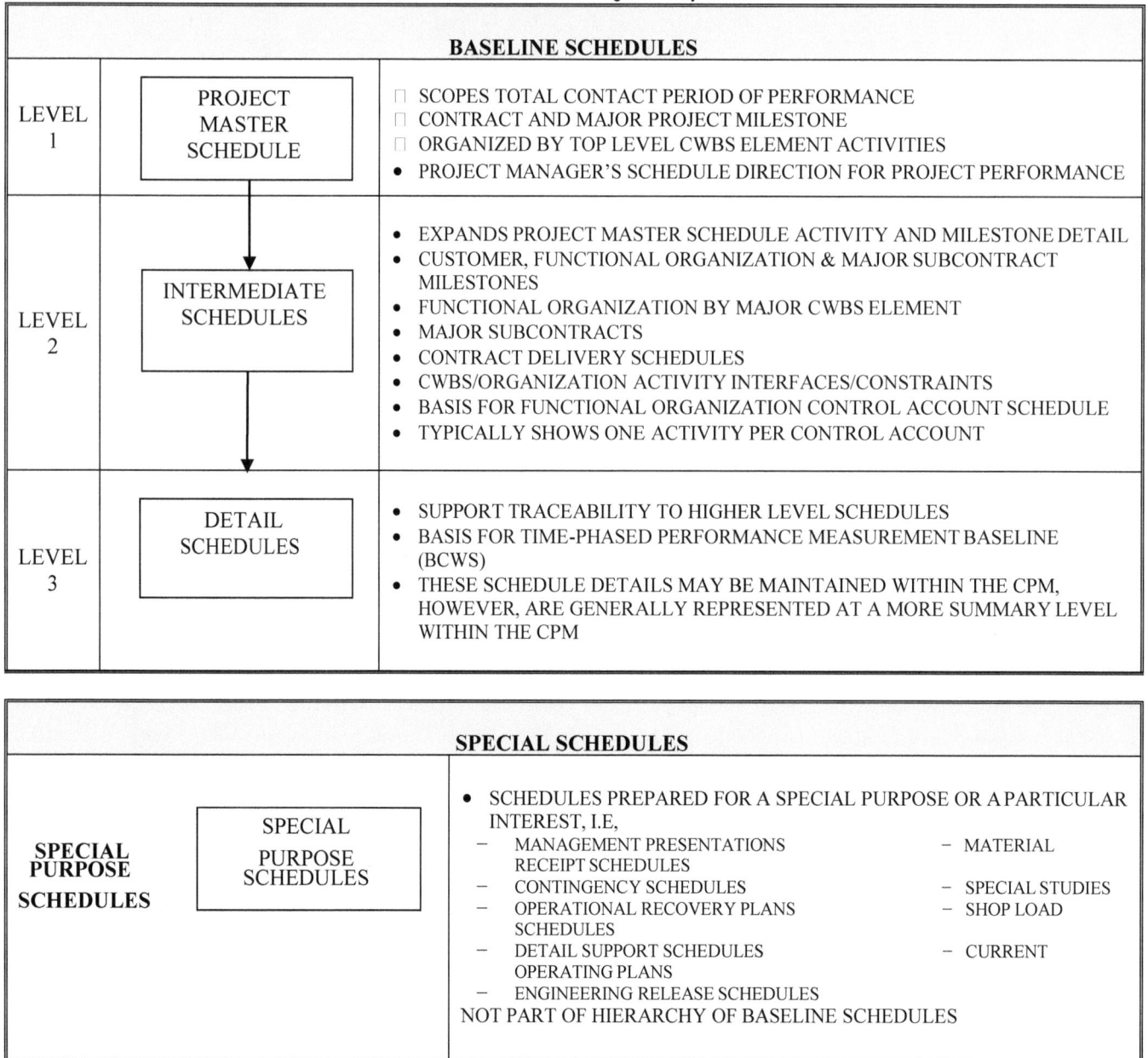

Figure 6-1   Schedule Hierarchy

					2011						2012													2013				

Activity ID	Start	Finish	Original Duration	Remaining Duration
PROJECT COORDINATION				
General Condtions				
E1010000	19-Aug-11 A		0	0
E1019100		16-Nov-11 A	0	0
E1019300		15-Nov-12	0	0
E1019500		30-Nov-12	0	0
E1019900		27-Feb-13	0	0
ENGINEERING				
General Condtions				
E1011000	19-Aug-11 A	16-Nov-11 A	80	0
PROCUREMENT				
General Condtions				
E1013000	10-Jan-12 A	15-Nov-12	220	63
CONSTRUCTION				
Civil and Existing Conditions				
E1025000	10-Dec-11 A	03-Apr-12 A	90	0
Concrete				
E1030000	10-Feb-12 A	28-Jun-12 A	100	0
Process Gas & Liquid Handling, Purifica...				
E1430000	04-Apr-12 A	30-Nov-12	170	72
START UP				
Process Gas & Liquid Handling, Purifica...				
E1430010	03-Dec-12	27-Feb-13	60	60

Milestones and bars:
- Contract Award
- Engineering Complete
- Procurement Complete
- Construction Complete
- Substantial Complete
- Design Engineering
- General Procurement
- Civil Construction
- Concrete Construction
- Process Construction
- Commissioning and Start Up

Legend:
- ▬ Actual Work
- ▬ Critical Remaining Work
- ▭ Remaining Work ◆ ◆ Milestone

Page 1 of 1

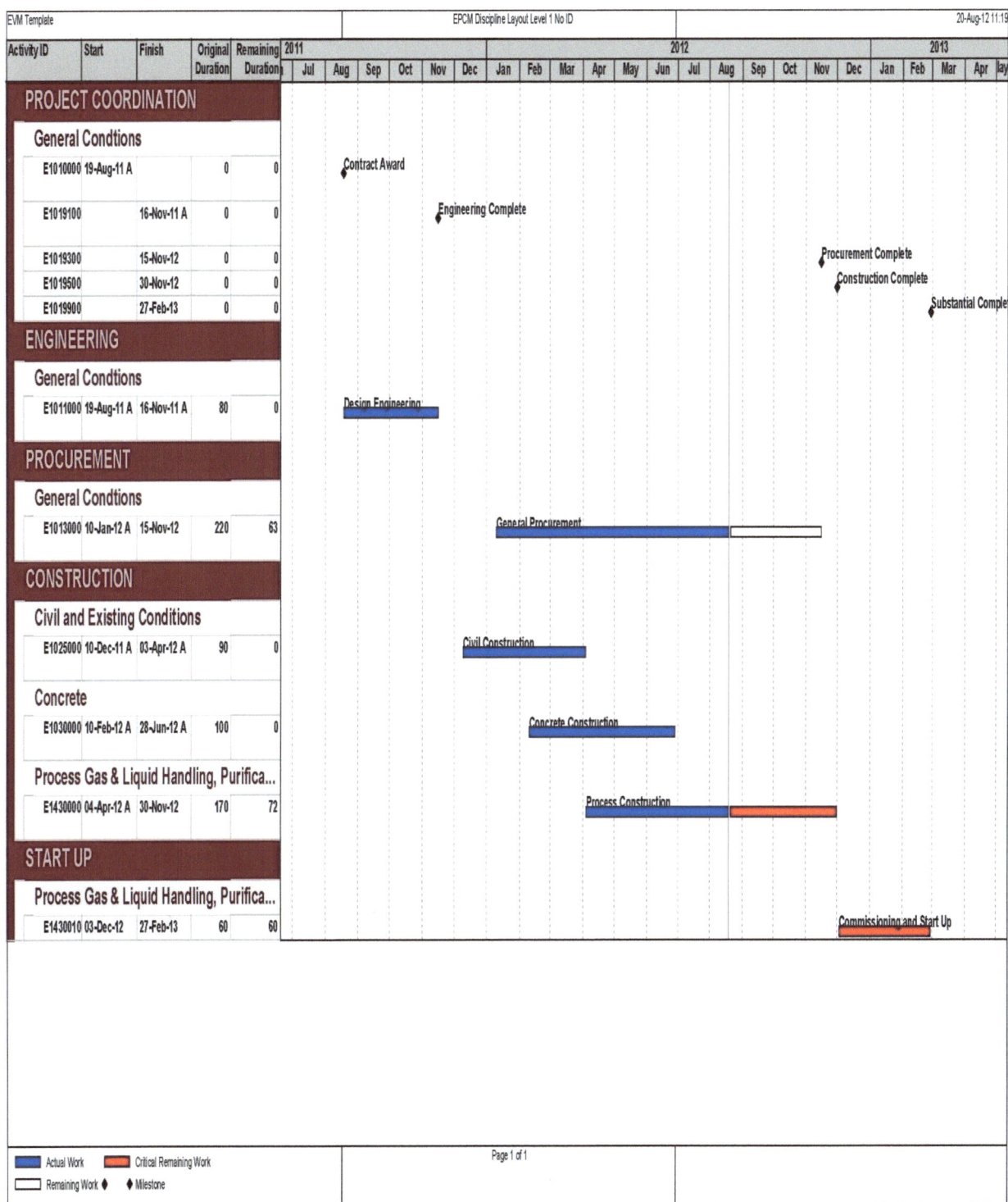

Figure 6-2   Sample Project Master Schedule

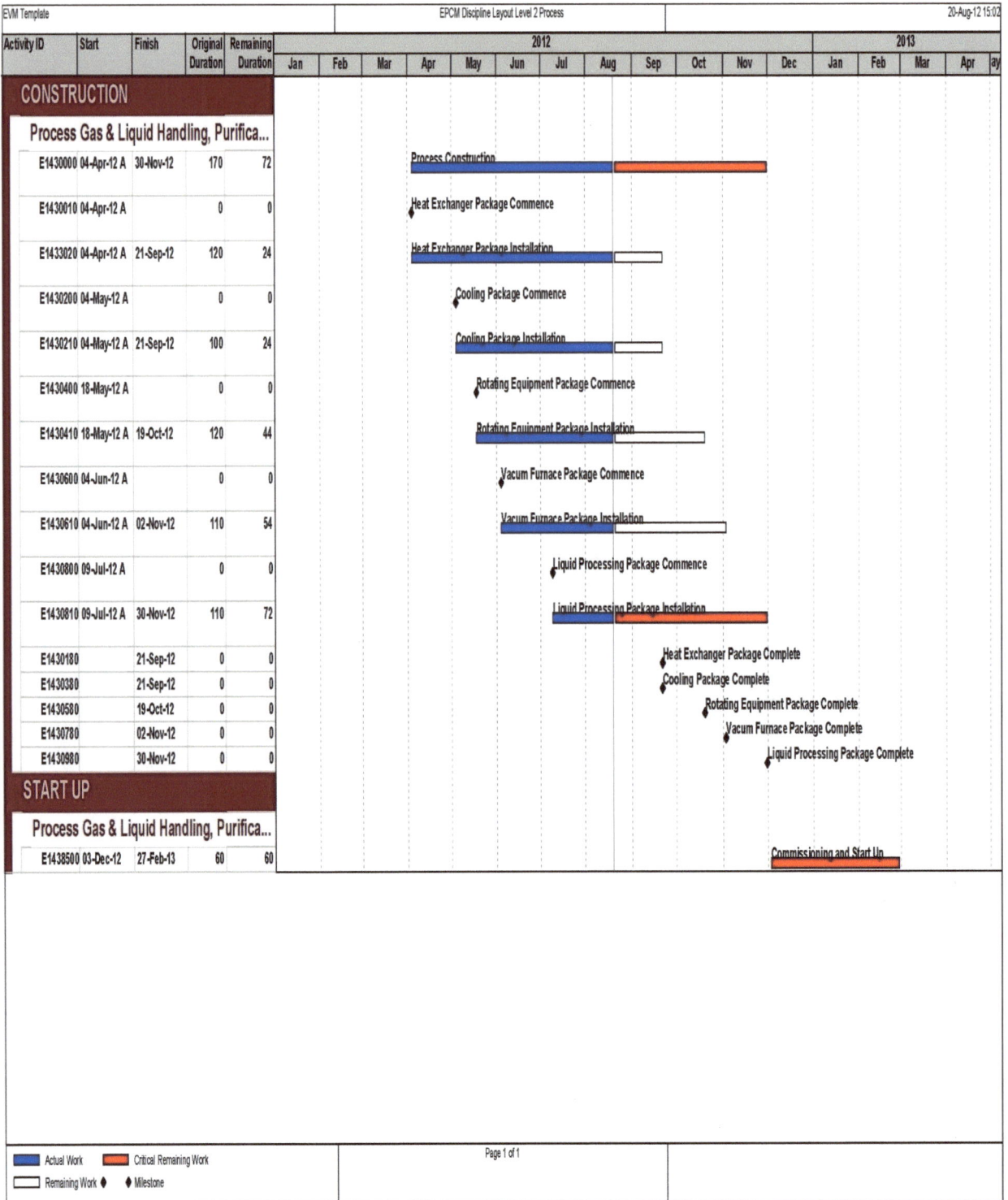

Figure 6-3 Sample Intermediate Schedule

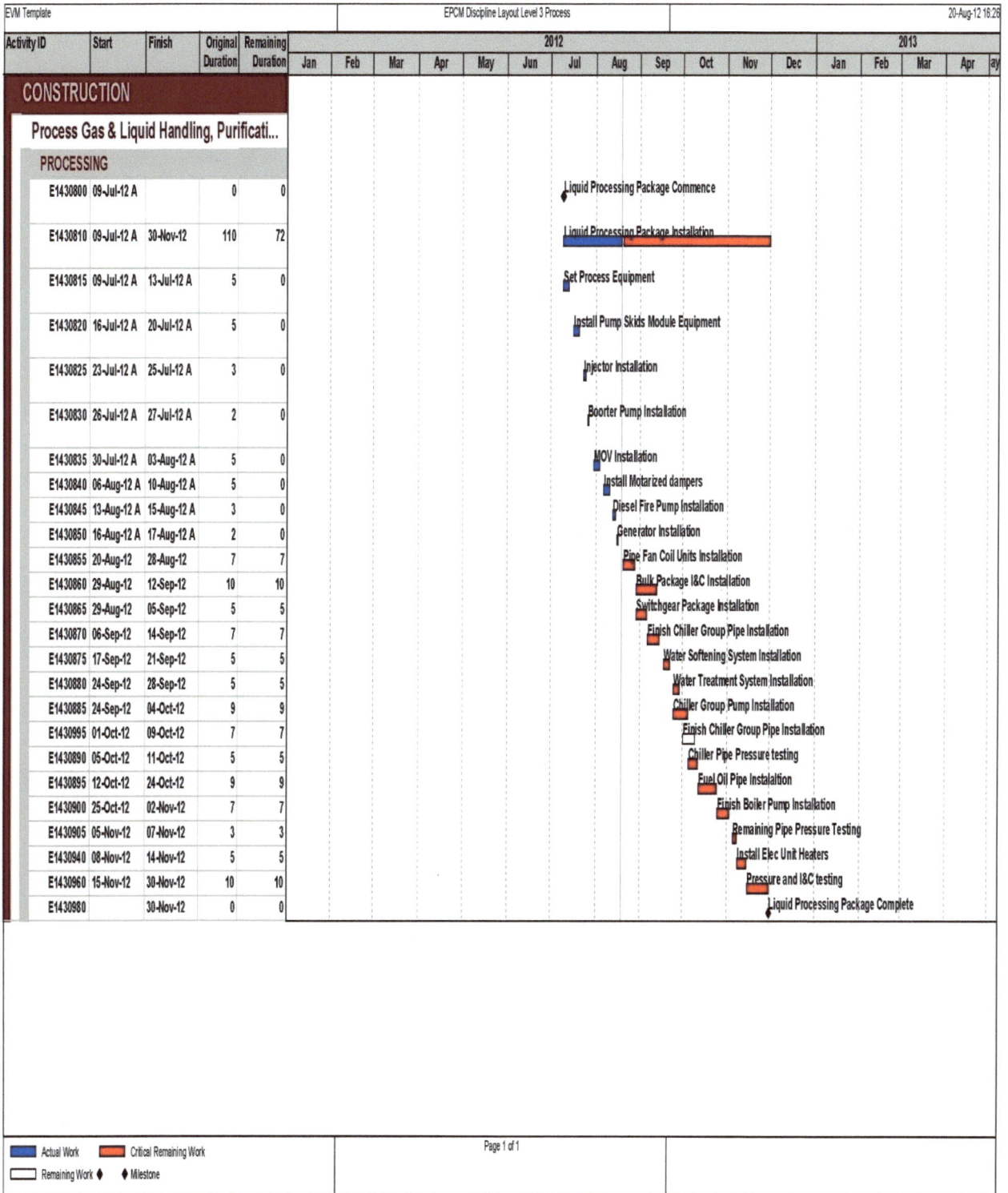

Figure 6-4   Sample Detailed Schedule

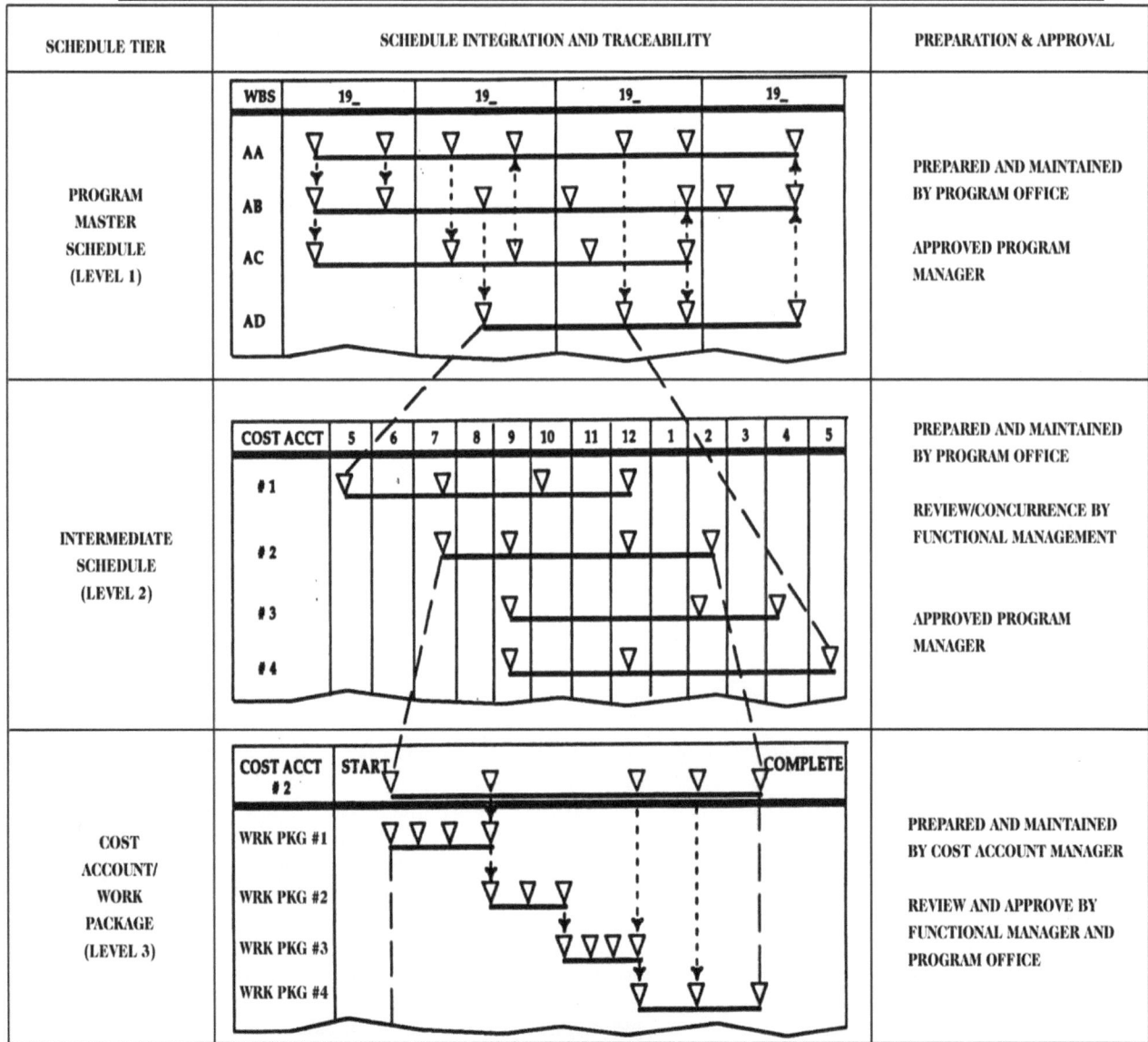

Figure 6-5   Schedule Integration and Traceability

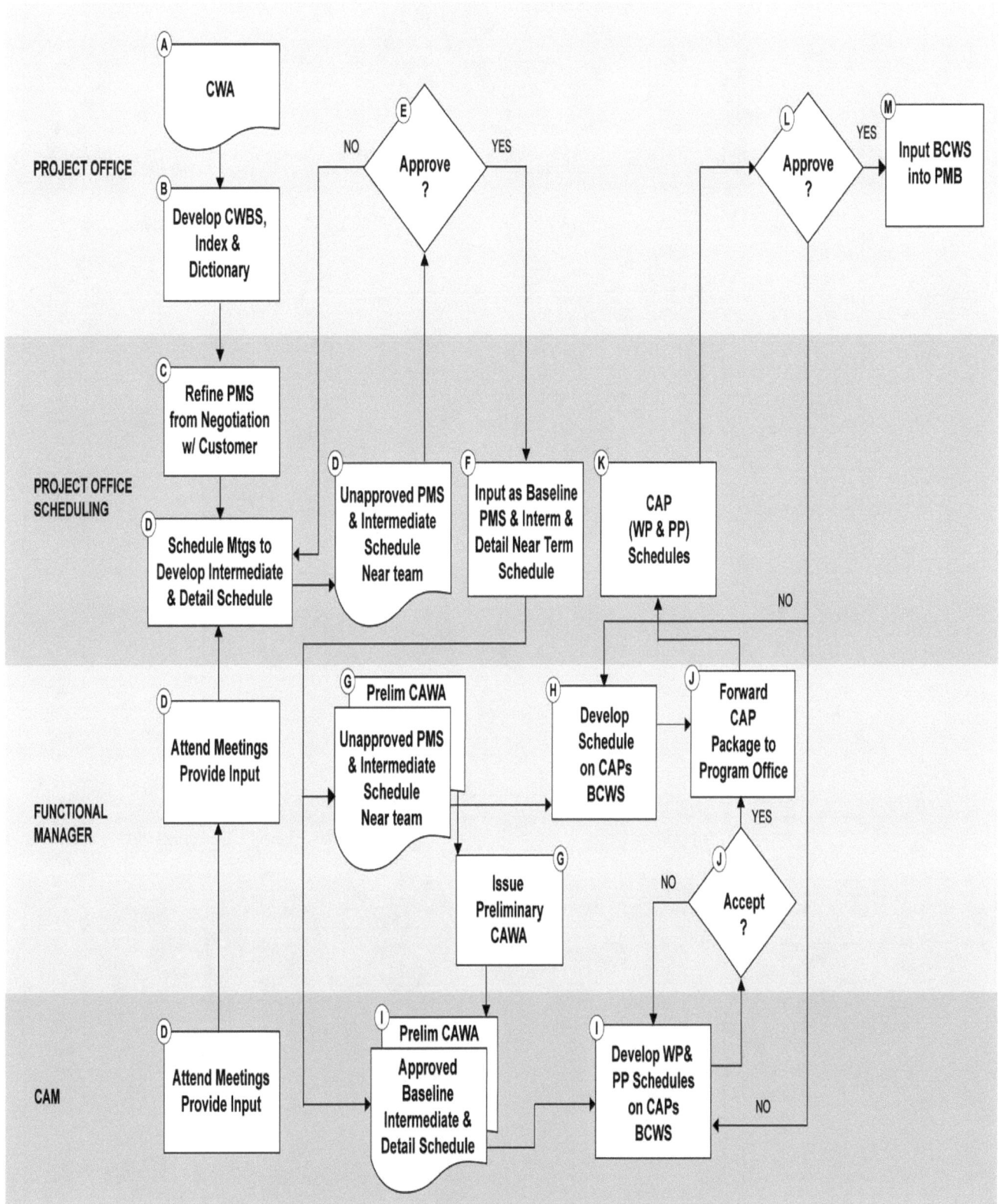

Figure 6-6   Schedule Development Flowchart

Use or disclosure of data contained on this sheet is
subject to the restriction on the title page of this document.

Page 59

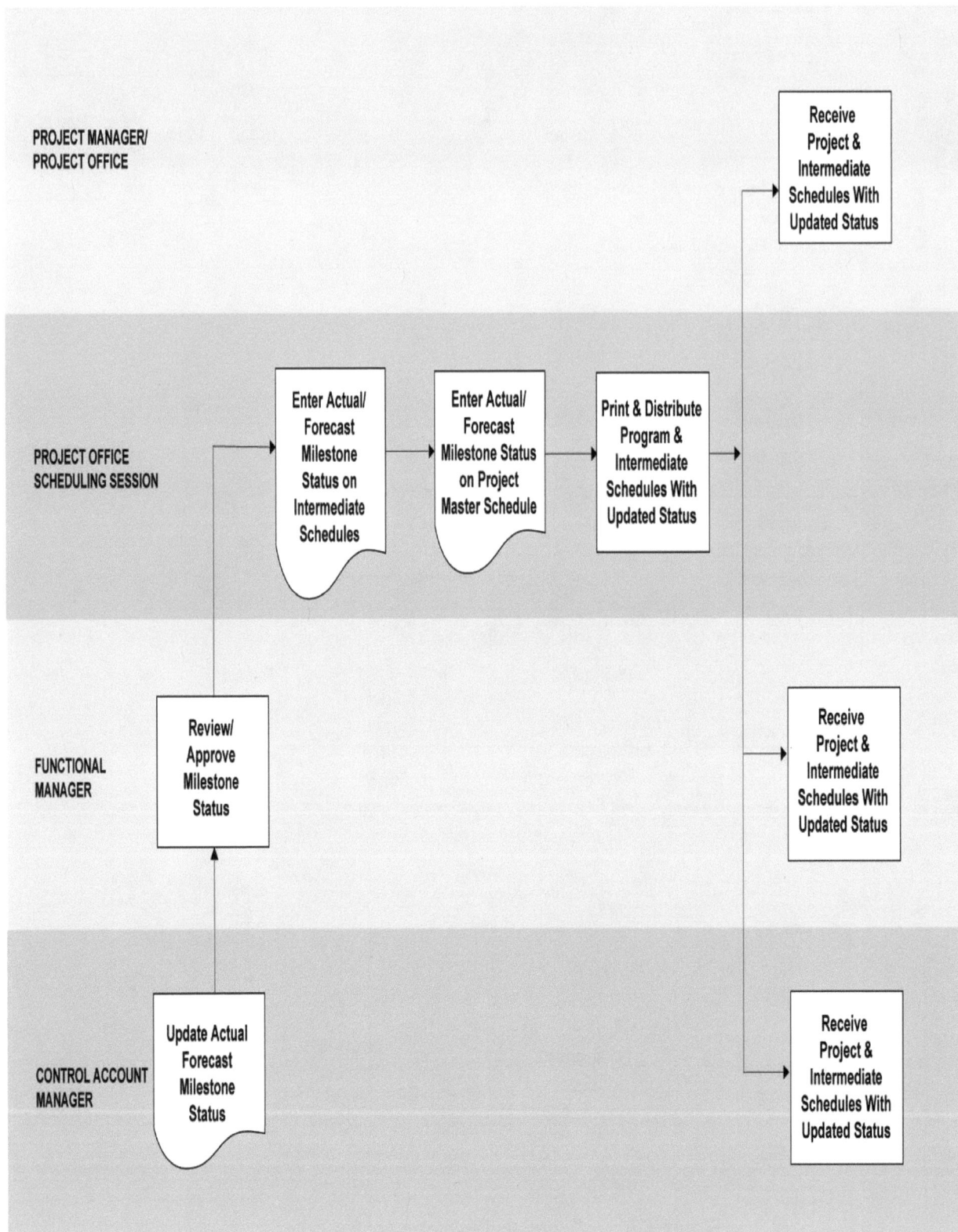

Figure 6-7   Schedule Status Flowchart

Use or disclosure of data contained on this sheet is
subject to the restriction on the title page of this document.

Page 60

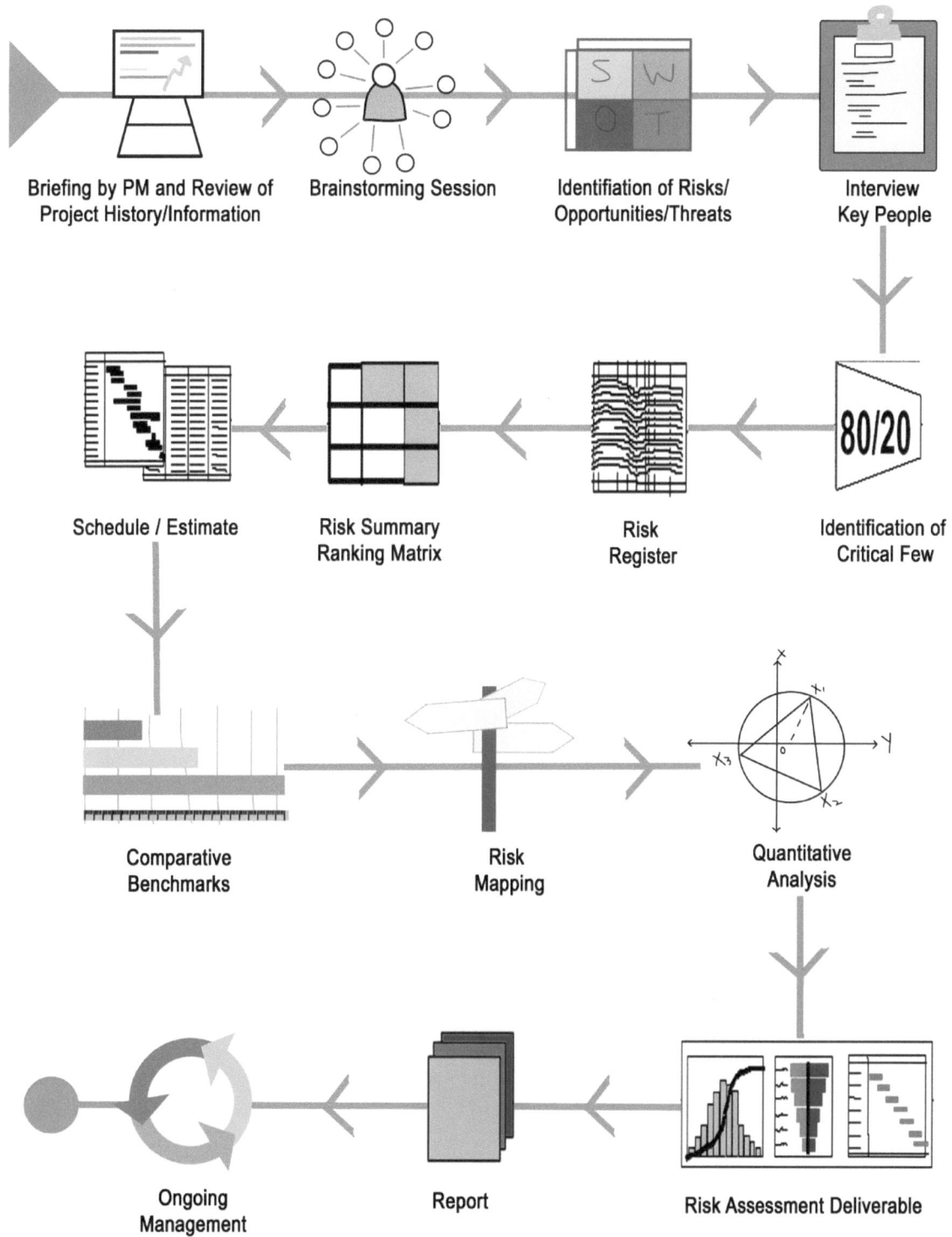

Figure 6-8   **Project Risk Analysis Process Diagram**

Use or disclosure of data contained on this sheet is
subject to the restriction on the title page of this document.

Page 61

7. Work Authorization

Use or disclosure of data contained on this sheet is
subject to the restriction on the title page of this document.

Page 62

7.1 Overview

The work authorization process (Figure 7-1) is used to document and authorize the SOW, schedule, and budget. The process, which implements work direction to organizations supporting the project, including subcontractors, begins with the customer's authorization in the form of a contract, letter, or change authorization. The customer provides Business Development/Contracts, which in turn provides the Project Manager, with authority to proceed and funding limitations via a CWA. The Project Manager then releases budget and work authorization to Functional Managers via a PWA. After receiving the PWA, each Functional Manager issues a CAWA down to the performing CAM providing budget and authorization to proceed. Budget logs track the issuance and receipt of budgets.

7.2 Work Authorization

No work is authorized to proceed without a properly executed work authorization. During contract performance, work authorizations are issued or changed under the following circumstances:

☐ Adding or deleting scope of work, budget, and schedule from an existing work authorization

☐ Transferring scope of work, budget, and schedule from one organization to another

☐ Issuing budgets from the UB

☐ Issuing MR

☐ Incorporating changes in authorized funding

7.2.1 Contract Work Authorization

When LGM INTERNATIONAL receives a contract, Business Development/Contracts issues a CWA (Figure 7-2) authorizing the Project Manager to prepare and release PWAs to the appropriate functional organizations. The CWA includes reference to the contract authorization, contract line item identification, schedule reference, contract negotiation summary, and funding limitations if applicable.

To initiate early post-award work by functional organizations prior to internal negotiations of budgets at the Control Account level, initial work authorizations are often needed. The initial work authorizations contain budgets and task descriptions necessary to initiate contractual work on a timely basis.

7.2.2 Project Work Authorization

The PWA (Figure 7-3) enables the Project Manager to authorize the initiation and stoppage of work. The PWA establishes authorized budgets for hours and direct labor, material, subcontract, overhead, general and administrative, and ODCs. Budget planning is performed within the budgets and schedules specified on the work authorization documents.

The PWA is the vehicle by which responsibility for budget, schedule, and task requirements is delegated from the Project Manager to Functional Managers. The PWA represents a process of negotiation and commitment. The signing of a PWA by all parties (the Project Manager and Functional Manager) represents a bilateral commitment to authorize and manage the work within the budget and schedule agreed on. The PWA continues to be the control document at the functional level unless modified by an agreement.

Work is not authorized and charges cannot be accumulated against contract activity until the PWA has been executed by the Project Office and a CAWA has been executed by the Functional Manager.

7.2.3 Control Account Work Authorization

At the Control Account level, the CAWA serves the same purpose as the PWA at a lower level. When signed by all parties (Functional Manager and CAM), it represents a commitment to perform and manage the work within the agreed-on budget and schedule.

Use or disclosure of data contained on this sheet is subject to the restriction on the title page of this document.

Page 63

The Functional Manager authorizes work to the CAMs within the authority provided in the PWA via a CAWA (Figure 7-4). The CAWA includes relationship to the CWBS element, responsible organization identification, Control Account scope of work, schedule, and budget. The CAWA budget is identified by direct labor hour and labor dollars and/or material dollars, subcontract dollars, ODC dollars, and overhead/G&A.

The CAM establishes a CAP (Figure 7-5) for each Control Account assigned to him or her by the Functional Manager. The CAP contains Control Account identification, schedule, responsibility, budget by dollars and element of cost, time-phased planning, milestone planning, a description of the discrete work to be performed/charged to the Work Package, and a performance measurement methodology. Work Packages are normally planned progressively through the project life defining discrete work for the next 6 months. As additional Control Account effort is planned in detail, new Work Packages are created. CAMs may assign Work Package Managers to perform day-to-day management of specific Work Packages within the Control Account.

With completion of CAPs for all Control Accounts, total authorized work is released to the performing organizations based on approved CAPs.

7.3   Work Authorization Flowchart

Figure 7-6 provides a comprehensive flowchart of the work authorization process and incorporates the budgeting and scheduling subsystems. The process is described in the following paragraphs.

A. Upon receipt of a contract, Business Development/Contracts prepares and releases a Contract

Work Authorization (CWA). The CWA authorizes the Project Manager to expend resources to accomplish the contract SOW. It also includes contract line item identification and schedule references and establishes the CTC.

B. Upon receipt of the CWA, the Project Office is responsible for accomplishing the initial contract planning. This planning involves defining the contract SOW and establishing initial budget and schedule baselines, as described in Steps C through I.

C. The Project Office develops the CBB log, placing all budget in UB.

D. The Project Manager reviews contract risk areas and identifies an MR value. The transfer of budget to MR is documented in the CBB log, identifying the value remaining to be distributed to the functional areas and ultimately the Control Accounts.

E. Simultaneously, the Project Office refines the proposed CWBS or extends the RFP WBS as appropriate and the associated CWBS Index and Dictionary. This begins the work authorization process as the contract SOW is broken down into manageable pieces of effort.

F. Based on the CWBS and the CWBS Index and Dictionary, the Project Office develops a preliminary RAM. At this point, the RAM may identify effort to a major functional organization only, not to a specific lower level department or CAM. Budgets are identified to each appropriate CWBS/functional organization intersection.

G. Concurrent with the development of the preliminary RAM and budget identification, the Project Office develops the PMS or modifies the proposed PMS based on contract requirements, major milestones, and deliverables.

H. Based on the CWBS and the PMS, the Project Office develops preliminary Intermediate Schedules, which displays significant activities and their interdependencies. Preliminary Intermediate Schedules are approved by the Project Manager.

I. The Project Office develops and the Project Manager issues preliminary PWAs to the Functional Managers. The budgets on these preliminary PWAs are based on the preliminary budget inputs by the Functional Managers, after any changes resulting from negotiations have been considered. The PWA schedules are based on the PMS and preliminary Intermediate Schedules. The SOW on the PWAs is based on the CWBS Dictionary. The PMSs, Intermediate Schedules, CWBS, and CWBS Index and Dictionary are also issued to the Functional Managers.

J. The Functional Managers receive the preliminary PWAs, schedules, and CWBS. Based on the tasks they are responsible for, they select appropriate CAMs from within their organization. If Functional Managers' efforts are large or they have many CAMs, they might develop a RAM for their functional organization. This document helps identify CAMs for specific pieces of effort and relate them to areas of the CWBS. It also enables the identification of budgets to specific tasks. Functional Managers who have small efforts or just a few CAMs may find a functional level RAM unnecessary.

K. The Functional Managers develop preliminary CAWAs outlining a basic SOW and identifying an overall schedule and budget.

L. The Functional Managers issue the preliminary CAWAs, along with copies of the CWBS, CWBS Index and Dictionary, and PMS and Intermediate Schedules.

M. Each CAM receives the preliminary CAWAs for the effort for which he or she is responsible.

N. Upon receiving the preliminary CAWAs and supporting information, the CAMs develop Cost

Account Plans (CAPs). This includes defining tasks, breaking them into smaller Work Packages, scheduling the Work Packages, and time phasing the resources necessary to accomplish the tasks. CAMs use the preliminary CAWA schedule and budget as a guideline in detail planning but are not constrained by them if either proves to be unrealistic.

O. The CAMs review their completed detail planning with their Functional Managers. Any areas of concern (i.e., definition of scope of work, schedule, and/or budget for specific tasks) are resolved between the CAMs and Functional Manager before they meet with the Project Office. The CAMs and Functional Manager interface with the Project Controls Scheduling Section to understand schedule implications of proposed CAPs.

P. After the Functional Manager reviews and agrees with the CAPs prepared by the CAMs, the Functional Manager presents his or her position to the Project Manager using the CAPs (as negotiated with the CAMs). This involves the scope of work, budget requirements, and schedule plan on the proposed PWA form.

Q. The proposed PWA, with backup package, is presented to the Project Manager. The Project

Manager reviews the positions presented by the Functional Managers. The Project Manager is supported by his or her staff in discussions with the functional organizations, specifically the scheduling and cost control organizations of Project Controls. There may be a negotiation process because the Project Manager has an obligation to meet all contract requirements: scope of work, schedule, and budget.

R. After the Project Manager negotiates with all Functional Managers, the initial contract planning is revised as necessary. This is described in Steps S through V.

S. The CWBS and CWBS Index and Dictionary are reviewed and revised if necessary. Changes to these documents are normally minor and reflect a more accurate understanding of the contract SOW or slight alterations in the lowest levels of the CWBS. At this point, the CWBS and CWBS Index and Dictionary are put in their final form and delivered to the customer if required by contract.

T. The RAM, which identifies effort to specific CAMs, is updated.

U. The PMS is revised if necessary. Again, changes to the PMS are normally minor because it reflects contract delivery dates and major milestones.

V. The Scheduling Organization makes any necessary revisions to the preliminary Intermediate Schedules. Changes that do not support the PMS are not made. Both the preliminary and the final Intermediate Schedules support the PMS. The Intermediate Schedules ultimately provide a logical tie between the PMS and the Control Account Schedules.

W. Upon completion of planning, the Project Office prepares the final PWAs. These documents identify the scope of work, budget, and schedule that have been agreed on as being the responsibility of specific Functional Managers. The final PWA is approved by the Project Manager.

X. The Functional Managers receive the final PWAs and copies of the final CWBS, CWBS Index and Dictionary, and PMS and Intermediate Schedules.

Y. The Functional Managers provide final guidance to CAMs to iterate their Control Account planning based on the negotiations that occurred with the Project Manager, PWAs, and other documents received from the Project Office.

Z. The CAMs iterate their CAPs, if necessary, to accomplish the Control Account scope of work within the negotiated schedule and budget.

Use or disclosure of data contained on this sheet is subject to the restriction on the title page of this document.

Page 65

AA. The Functional Manager authorizes, by signature, the CAMs to perform the specified scope of work within a specific budget and an established schedule via a final CAWA.

BB. The Functional Manager forwards copies of approved CAPs and CAWAs to the Project Office.

CC. The Project Office receives copies of the CAPs and CAWAs. These are input into the LGM INTERNATIONAL EVMS database to set BCWS and adjust the CBB, UB, and MR logs. The CAWA log is updated (Figure 7-7).

DD. The CAMs receive copies of their final CAWAs.

EE. The CAMs begin work in accordance with the schedules identified in the CAPs.

Use or disclosure of data contained on this sheet is
subject to the restriction on the title page of this document.

Page 66

DOCUMENT	PREPARED BY	APPROVED BY	ISSUED TO	PURPOSE
CONTRACT	CUSTOMER	CUSTOMER PROCURING CONTRACTING OFFICER	LGM INTERNATIONAL	• Define contract scope of work, period of performance, schedule for deliverables, contract schedule milestones, technical specifications, data requirements, contract price, target cost, fee funding. • Provide authority to proceed with work.
CONTRACT WORK AUTHORIZATION (CWA)	BUSINESS DEVELOPMENT/ CONTRACTS	BUSINESS DEVELOPMENT/ CONTRACTS AS AGENT FOR BUSINESS UNIT PRESIDENT	PROJECT MANAGER	• Provide authority to proceed. • Identify contract number, assign internal project number, and define contract performance requirements. • Authorize/issue target cost to Project Manager as Contract Budget Base.
PROJECT WORK AUTHORIZATION (PWA)	PROJECT OFFICE	PROJECT MANAGER	FUNCTIONAL MANAGERS	• Define/assign scope of work to functional organization managers. • Identify schedule period of performance for assigned work • Issue budget and authorize initiation of baseline planning.
CONTROL ACCOUNT WORK AUTHORIZATION (CAWA)	FUNCTIONAL ORGANIZATION	FUNCTIONAL MANAGER	CONTROL ACCOUNT MANAGER	• Define/assign scope of work to Control Account Manager. • Authorize development of CAP. • Identify budget and schedule constraints. • Authorize scope of work, budget, schedule and performance of work.
CONTROL ACCOUNT PLAN (CAP)	CONTROL ACCOUNT MANAGER	FUNCTIONAL MANAGER AND PROJECT OFFICE	CONTROL ACCOUNT MANAGER	• Identify control account, planning package and work package scope of work, schedules, time phased budgets, resource types and CAM.
PURCHASE REQUEST (PR)	FUNCTIONAL MANAGER/ CONTROL ACCOUNT MANAGER	PROJECT MANAGER/ FUNCTIONAL MANAGER/ CONTROL ACCOUNT MANAGER AS APPROPRIATE	PROCUREMENT/ MATERIAL DEPARTMENT MANAGER	• Define products and services to be procured, delivery schedules or period of performance, applicable technical specifications, and budgeted value and cost collection number(s). • Authorized placement of purchase order or subcontract.
SUBCONTRACT	SUBCONTRACT ADMINISTRATOR	CORPORATE PROCUREMENT PROJECT MANGER/ FUNCTIONAL MANAGER AND/OR PROCURMENT/ MATERIAL MGR AS APPROPORIATE	SELECTED SUBCONTRACTOR	• Define products to be delivered and services to be performed, delivery schedules or period of performance, applicable specifications, price target cost, fee, CLIN(s) and contract type. • Authorize subcontractor work performance.
PURCHSE ORDER (PO)	PROCUREMENT/ MATERIAL MANAGER	CORPORATE PROCUREMENT PROJECT MANGER/ FUNCTIONAL MANAGER AND/OR PROCURMENT/ MATERIAL MGR AS APPROPORIATE	SUPPLIER/VENDOR	• Define products to be delivered and services to be performed, delivery schedules or period or performance, applicable specifications, quantities, unit price and total purchase order value. • Authorize vendor/supplier to deliver products and/or perform services.

Figure 7-1   Work Authorization Process

CONTRACT WORK AUTHORIZATION (CWA)		CONTRACT NO.
TO PROJECT MANAGER		REV. NO.
PROJECT		CWA LOG #
CONTRACT, DRAWING, SPECIFICATION, EQUIPMENT NO., OR OTHER REFERNECES		DATE

DESCRIPTION, CONTRACT TYPE AND FUNDING INFORMATION

SCHEDULE	PRIOR APPROVED TOTAL PERIOD OF PERFORMANCE	PERIOD OF PERFORMANCE THIS CWA
	START:	START:
	STOP:	STOP:

COST	PRIOR TOTAL	
	THIS CWA	
	TOTAL	

APPROVALS	BUSINESS DEVELOPMENT/ CONTRACTS (AS AGENT FOR BUSINESS UNIT PRESIDENT) _____ SIGNATURE _____ DATE	ACCEPTANCE	PROJECT MANAGER: _____ SIGNATURE _____ DATE
		ACKNOWLEDGE	PROJECT CONTROLS MANAGER: _____ SIGNATURE _____ DATE

Figure 7-2   Sample Contract Work Authorization

Use or disclosure of data contained on this sheet is subject to the restriction on the title page of this document.

Page 68

PROJECT WORK (PWA)	AUTHORIZATION		PWA NO.	
			REV. NO.	
	DATE		CWA REF.	
			PWA LOG NO.	

TO FUNCTIONAL MANAGER | **CONTRACT/JOB NO.** | ☐ Preliminary ☐ Proposed ☐ Final ☐ Revision Request

PROJECT | **FUNCTIONAL AREA**

DRAWING, SPECIFICATIONS, EQUIP NO., OR OTHER REFERENCE | **CWBS REF. #(S)**

ORIGIN IF REVISION REQUEST

DESCRIPTION AND/OR SKETCH | ☐ SCOPE OF WORK ☐ REASON OF CHANGE

FUNCTIONAL MANAGER

SCHEDULE IMPACT

SCHEDULE REFERENCE(S)	PRIOR SCHEDULE	THIS PWA SCHEDULE
	START:	START:
	STOP:	STOP:

COST IMPACT

	MAN-HOURS	LABOR $	MATERIALS $	SUBCONT $	OTHER DIRECT COST $	OVERHEAD $	G&A $	TOTAL $
PRIOR TOTAL	OFFICE							
	FIELD							
THIS PWA	OFFICE							
	FIELD							
TOTAL	OFFICE							
	FIELD							

APPROVALS

PROJECT CONTROLS CONCURRENCE

_____ PROJECT CONTROLS MANAGER SIGNATURE
DATE
PRELIMINARY

PROJECT MANAGER SIGNATURE   DATE
PROPOSED/ REVISION REQUEST

FUNTIONAL MANAGER SIGNATURE   DATE
FINAL

PROJECT MANAGER SIGNATURE   DATE

FUNCTIONAL MANAGER ACCEPTANCE   DATE

Figure 7-3   Sample Project Work Authorization

CONTROL ACCOUNT WORK (CAWA)	AUTHORIZATION		CAWA NO.	
			REV. NO.	
	DATE		PWA REF.	
			CAWA LOG #	

TO CONTROL ACCOUNT MANAGER	CONTRACT/JOB NO.	☐ Preliminary ☐ Proposed ☐ Final ☐ Revision Request
PROJECT	FUNCTIONAL AREA	
DRAWING, SPECIFICATIONS, EQUIP NO., OR OTHER REFERENCE	CONTROL ACCT #	

	DESCRIPTION AND/OR SKETCH	☐ SCOPE OF WORK ☐ REASON OF CHANGE
ORIGIN IF REVISIO	FUNCTIONAL MANAGER	CONTROL ACCOUNT MANAGER

	SCHEDULE REFERENCE(S)	PRIOR SCHEDULE	THIS CAWA SCHEDULE
SCHEDULE IMPACT		START:	START:
		STOP:	STOP:

COST IMPACT		MAN-HOURS	LABOR $	MATERIALS $	SUBCONTR $	OTHER DIRECT COST $	OVERHEAD $	G&A $	TOTAL $
	PRIOR TOTAL	OFFICE							
		FIELD							
	THIS CAWA	OFFICE							
		FIELD							
	TOTAL	OFFICE							
		FIELD							

APPROVALS	PROJECT CONTROLS CONCURRENCE:
	_____ DATE PROJECT CONTROLS MANAGER SIGNATURE
	PRELIMINARY _____ _____ FUNCTIONAL MANAGER DATE
	PROPOSED/ REVISION REQUEST _____ _____ CONTROL ACCOUNT MANAGER SIGNATURE DATE
	FINAL _____ _____ FUNCTIONAL MANAGER SIGNATURE DATE
	_____ _____ CONTROL ACCOUNT MANAGER ACCEPTANCE DATE

Figure 7-4   Sample Control Account Work Authorization

Report: CAP LJ	Control Account Planning Sheet									Page : 4	
Project: ANCDF	Description: PROJECT XYZ			Approval: Project Office							
Run Date: 08/03/03	Status Date: 07/31/03			Function Manager							
				Control Account Manager							
Control Account: 14313			Description: CHB CONCRETE						CA Manager: SMITH		
Scheduled Start: 07/15/03			Scheduled Finish: 05/25/04						Status: Open		

Work Package: 02	Description: CIVIL STRUCTURES				Scheduled Start: 07/15/03			Scheduled Finish: 05/25/04			
PMT: Milestone	Status: Open	Jul 03	Aug 03	Sep 03	Oct 03	Nov 03	Dec 03	Jan 04	Feb 04	Mar 04	Apr 04	May 04	At Complete
	BCWS	16	28	39	55	71	110	83	50	50	28	20	550
Cost Element	BCWP	50	0	0	0	0	0	)	0	0	0	0	
FND/CIVIL MATERIALS	ACWP	58	0	0	0	0	0	)	0	0	0	0	
	BCWS	14	23	36	49	66	102	76	45	45	25	19	500
	BCWP	45	0	0	0	0	0	)	0	0	0	0	
LABOR	ACWP	41	0	0	0	0	0	)	0	0	0	0	
	BCWS	30	51	75	104	137	212	159	95	95	53	39	1050
Work Package	BCWP	95	0	0	0	0	0	0	0	0	0	0	
Totals	ACWP	99	0	0	0	0	0	0	0	0	0	0	

Figure 7-5   Sample Control Account Plan

Use or disclosure of data contained on this sheet is
subject to the restriction on the title page of this document.

Page 71

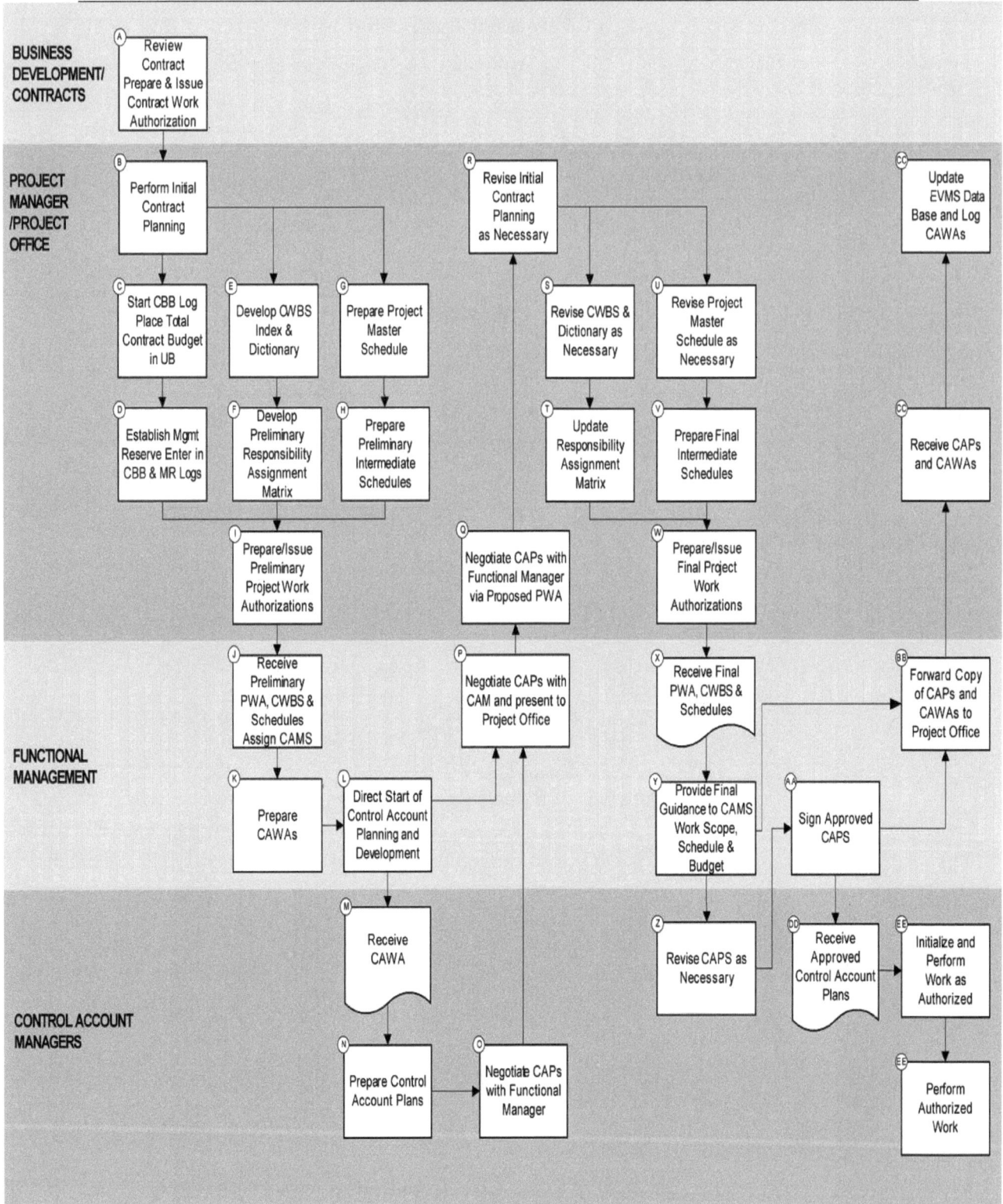

Figure 7-6   Work Authorization Flowchart

Use or disclosure of data contained on this sheet is
subject to the restriction on the title page of this document.

Page 72

Log No.	Rev. No.	PWA No.	Orgin Date	Description	Approval Date	BUDGET			SCHEDULE			
						Prior Total $	This Revn.	Total $	Prior Approved Dates		This Approval	
									Start	Stop	Start	Stop

Figure 7-7   Sample CAWA Log

8. Performance Measurement

Use or disclosure of data contained on this sheet is
subject to the restriction on the title page of this document.

Page 74

8.1   Overview

This section presents the various techniques used to establish the Performance Measurement Baseline (PMB), or BCWS, and the means for measuring completed work or BCWP. BCWP and BCWS are two key data elements of overall data analysis and reporting, which is discussed in Section 13. Wherever possible, earned value or BCWP is generated using discrete Work Packages which represent a specific measurable activity within a Control Account. The Control Account, along with its associated Work Packages and Planning Packages form part of the next higher Contract Work Breakdown Structure (CWBS) element. Alternate measurement techniques include LOE and Apportioned Effort.

8.2   Principal Characteristics

The following are characteristics of the LGM INTERNATIONAL EVMS performance measurement subsystem.

A. Measurement is made at the lowest organizational level performing the work within a Control Account.

B. Measurement of cost and schedule performance is made in a systematic manner consistent with the manner in which the work was scheduled and budgeted.

C. BCWP is determined in a manner consistent with the way BCWS was developed.

D. Retroactive changes to BCWS and BCWP are prohibited except in the event of error correction or normal accounting adjustments.

8.3   Development of Performance Measurement

Once budget is distributed to a Control Account, Work Packages are developed that contain budgets in terms of labor hours, dollars and/or rates for specific packages of work that will be planned and measured across the time span of that package. Using the previous example, a Work Package may be "Munitions Building - Concrete - Building Foundations." The intent here is to subdivide the work into packages that have a shorter duration and represent specific tasks that have a common complexity and budget basis. For instance, "Foundation" is a Work Package distinct from "Elevated Slabs." Each Work Package is part of the Concrete Control Account, but is budgeted and planned differently due to a difference in the type of budget resources required, and/or the type of work.

Once detailed planning is accomplished, detail time-phased budgets are established at the lowest applicable levels and expressed in terms of labor hours and/or dollars. The time required to accomplish detail planning varies based on the complexity of the contract and the amount of information available. However, it should be accomplished as soon after contract award as possible. Detail Work Package planning represents the bulk of the effort at the beginning of the contract.

When establishing the Control Account, information may not be available to plan all Work Packages. In this case, far-term Planning Packages are utilized. Planning Package time-phased BCWS is based on the scheduled accomplishment of the work rather than any earned value technique. At this point, the BCWS is the time-phased budget for the Work Package and Planning Packages. Their sum equals the budget for the Control Account.

Per the LGM INTERNATIONAL rolling wave planning process, Control Account Planning Packages are detailed to Work Packages. When completed, the sum of the assigned, time-phased budgets (BCWS) equals the Control Account budget.

During this development period, progress is monitored to ensure compatibility to: 1) planning and controlling schedule and cost, 2) identifying actual cost for completed work and work in progress, and 3) accurately reporting actual progress related to the plan.

8.4   Performance Measurement Techniques

The time-phased budget (BCWS) represents the Performance Measurement Baseline (PMB). Arriving at this baseline for the Control Account results from identifying the tasks, scheduling the tasks, identifying resources, time-phasing those resources and determining the appropriate earned value method to be used for that Work Package. One criteria in determining the earned value method is that the BCWS must be planned in the same manner that the BCWP will be earned. This will avoid unrealistic variances. LOE Work Packages may arise within a Control Account but should be kept to a minimum to avoid distortion of performance measurement and evaluation. Where used, LOE Work Packages are identified independent of discrete Work Packages within the same Control Account. In these cases, ACWP is collected separately for the LOE and discrete portions of the Control Account.

8.4.1   Discrete Effort

The following earned value methods are available to measure performance on discrete Work Packages.

8.4.1.1   *Percent Complete of Events*

This method is used when completion of a Work Package is calculated at an event level in combination with other events and sums to the percent complete of the Work Package. Under this method, percent complete and earned value are determined as a specific Work Package advances towards its scheduled end date. Each Work Package can be subdivided into distinct measurable events each contributing to the earned value for that Work Package at any point during its duration.

As an example, assume Work Packages exist in the concrete Control Account for foundations, slabs, elevated structures, and blast walls. Each of these Work Packages is a subset of the Control Account - Concrete. In addition, each Work Package contains work that is consistent in complexity and budget value. That is, foundations are poured to support a piece of equipment and as such, require strength characteristics unique to it. By comparison, elevated structures require special forming and pouring techniques that are unique only to elevated pours. By establishing Work Packages in this manner, progress and earned value can be calculated and reported in a consistent manner against like activities.

Calculating progress at this Work Package level, however, is not in sufficient detail for timely control of the activity. Thus, the foundation Work Package is detailed, for progress purposes only, to the individual foundations to be poured (i.e., Foundation A, B, C, and D). It is at this level that earned value is determined.

Within each of these foundations distinct events take place which contribute to the completion and are measurable. Build formwork, install rebar, pour concrete, remove forms are examples of these events. Each event has a value, in terms of percent, that it contributes to completion. As the CAM verifies completion of these events, their percentage value is multiplied by the quantity of work for that task - in this case, cubic yards of concrete for Foundation A. This earned quantity for Foundation A is summed with the earned quantities for Foundation B, C, and D to arrive at percent complete, in terms of earned quantity, for foundations.

That earned quantity is multiplied by the budgeted unit cost per quantity. The result represents the BCWP for the reporting period. The same exercise is completed for the other concrete Work Packages and the sum of the Work The package BCWP equals the BCWP for the Control Account. At this point, comparisons and analysis can be made among BCWS, BCWP, and ACWP.

When the BCWS is developed for the Work Package, the same process is followed. The completion of each Work Package is scheduled and the budget is spread over the duration based on planned completion of the events.

Figure 8-1 illustrates the Control Account, Work Package, activity, and event hierarchical relationships.

8.4.1.2   *0/100 Technique*

This earned value method is sometimes used when a Work Package is scheduled to start and complete within the same accounting month. Earned value is achieved only upon completion of the entire Work Package.

8.4.1.3   *Direct Percent Complete*

When discrete Work Packages exist, and do not lend themselves to other measurable methods, the Direct Percent Complete method can be used. This technique allows the CAM to determine the percent complete based on visual inspection of work in place. However, the percent complete cannot exceed 80% of the total value until the task is 100% complete. This method is limited to Work Packages of a relatively short duration.

An example of the percent complete method is a fixed price subcontractor who will report progress based on the amount of scheduled activity completed that period. Verification is by physical and/or visual inspection of the tasks claimed as complete for that reporting period. The schedule for that subcontracted work is used as the document to note the completed work. An agreed upon percent complete is derived and the subcontractor's schedule is progressed.

The CAM maintains a record of the percent complete methodology that will be used for each Direct Percent Complete Work Package.

8.4.1.4   Direct Input

This method is used on subcontracted work where the subcontractor has CPR or C/SSR reporting responsibility to LGM INTERNATIONAL. In this instance, the BCWS and BCWP would be entered for the Control Account from the CPR or C/SSR.

Subcontractors who cannot support an EVMS program, based on agreement from surveillance team and client, will provide information in progress reporting in format as required to meet EVMS criterion.

8.4.1.5   Percent Start/Percent Complete Technique

This technique is used for Work Packages that are scheduled to start and complete within two accounting months. Two milestones are established; one for starting the Work Package and another to be claimed at completion. Each milestone carries a predetermined weight. BCWP is then taken in the same proportion as planned when the Work Package is started and again when it is completed.

8.4.1.6   Units Complete

This technique is used where a physical, measurable quantity exists and each quantity carries the same budget weight. BCWS is determined based on scheduled number of units each reporting period and BCWP is determined based on the actual number of units completed. Each is then multiplied by the budget value for each unit.

8.4.1.7   Milestone Achieved

This technique is used where output is to be measured by a series of interdependent milestones each of which denotes various stages of work completion. This method is used to minimize subjective assessment of status and provide objective assessment of work accomplishment. BCWS is time-phased to correlate to each milestone and BCWP is earned as the budgeted value of the milestone achieved.

8.4.2   Apportioned Effort

An Apportioned Effort is a budgeted, planned activity that is in support of one or more discrete Work Packages. Quality assurance and inspection are common apportioned tasks.

For example, when installing pipe, every fifth weld is x-rayed and inspected to ensure weld quality. Each x-ray and inspection requires two labor hours. This is an Apportioned Effort in support of the piping Work Package. BCWS is planned as a predetermined percentage of the discrete Work Package budgets in the same months. As earned value (BCWP) is determined each month for the discrete Work Packages, the same percentage is multiplied times the discrete Work Package (BCWP) to arrive at the Apportioned Effort Work Package (BCWP).

8.4.3   Level of Effort

LOE is used for activities whose only measurement criterion is the passage of time. LOE tasks are those which cannot be Work Packaged or apportioned and have no identifiable end product. Management activities are an example of an LOE. These budgets are planned based on the number of reporting periods and the level of intensity of this activity throughout the scheduled duration. These budgets are time phased by element of cost. BCWP is equal to BCWS at all times during the duration of the LOE task.

Use or disclosure of data contained on this sheet is
subject to the restriction on the title page of this document.

Page 77

Figure 8-2 recaps the earned value methods.

8.5   Control Account Plan

The CAP represents the entire product for establishing, planning, budgeting, and statusing the work currently authorized within the CAWA. It is the CAM's plan for accomplishing the work under his/her responsibility.

Once the Control Account Work Packages are defined by the CAM, it is the CAM's responsibility to develop the CAP in accordance with the Work Authorization Process described in Section 7. Inputs to this planning process might be equipment deliveries, drawing issues, restraints caused by interference by other activities, etc. Typically, the logic diagrams and the baseline schedules are developed in conjunction with the CAM, when available (see Section 6). The CAM begins planning the start and finish dates for each of the Work Packages within the Control Account to support the PMS and Intermediate Schedules. While these schedules show the broad activities to be accomplished, the CAM must place the more detail tasks (Work Packages) into the proper sequence. This is an iterative process since the Intermediate Schedule may need adjustments based on Control Account detail planning. The CAM must assign the Work Package budget across the duration of the Work Packages. This can be done based upon complexity of the work, personnel intensity, availability, cost of material, etc. Once completed, this time-phased budget for the Work Package is input into LGM INTERNATIONAL EVMS as the BCWS.

The resulting CAP is reviewed, signed, and dated by the CAM and Functional Manager in accordance with Work Authorization. Once finalized, the CAP documents the BCWS for each Work Package and Planning Package within the Control Account. It becomes the basis against which BCWP is compared for schedule performance. BCWP is the same budget value as the BCWS for that work completed. ACWP is compared with the BCWP for cost performance. Figure 8-3 presents an example of a CAP.

Use or disclosure of data contained on this sheet is subject to the restriction on the title page of this document.

Page 78

COST ACCOUNT
BCWS &BCWP SUMMED
ACWP DETERMINED

| STEEL | CONCRETE | PIPE | MECHANICAL |

WORK PACKAGE
BCWS &BCWP DETERMINED

| BLDG.SLABS | FOUNDATIONS | ELEVATED | BLAST WALLS |

ACTIVITY
PERCENT COMPLETE
DETERMINED

| FOUNDATION A | FOUNDATION B | FOUNDATION C | FOUNDATION D |

EVENTS
PROGRESS EARNED

INSTALL FORM

INSTALL REBAR

CONCRETE

REMOVE FORM

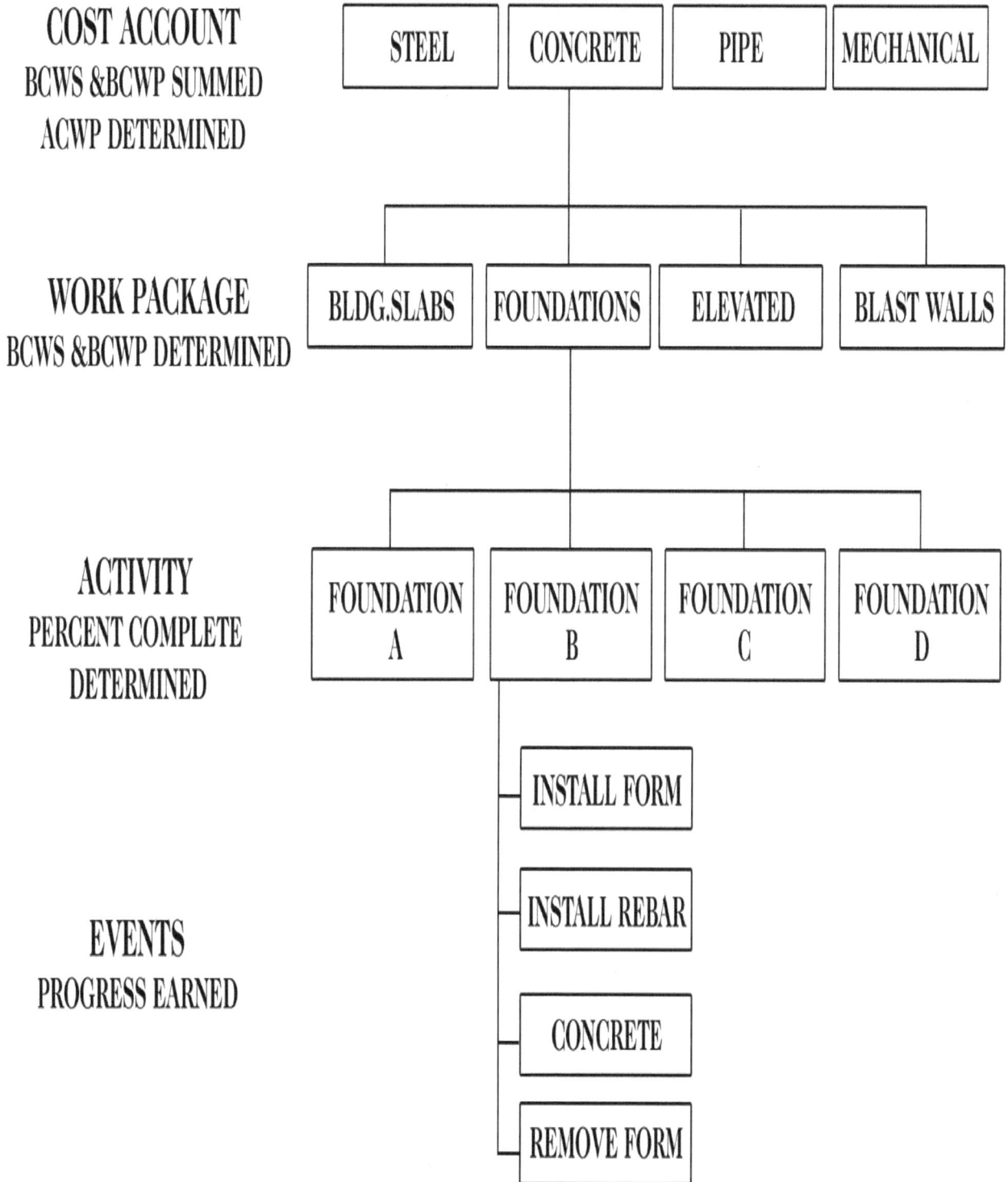

Figure 8-1   Sample Work Breakdown Structure Hierarchical Relationships

Use or disclosure of data contained on this sheet is
subject to the restriction on the title page of this document.

Page 79

EARNED VALUE (BCWP) CALCULATION METHODS

WORK PACKAGE PLANNING METHOD	USUAL TIME SPAN	BCWP CALCULATION
MEASURED WORK PACKAGES		
0/100	WITHIN 1 ACCOUNTING MONTH	100% OF BUDGET EARNED WHEN WORK PACKAGE IS COMPLETED
PERCENT START/ PERCENT COMPLETE	WITHIN 2 ACCOUNTING MONTHS	SET % OF BUDGET EARNED WHEN WORK PACKAGE IS STARTED, SET % EARNED WHEN WORK PACKAGE IS COMPLETED
PERCENT COMPLETE OF EVENTS	VARIABLE	BUDGET EARNED AS TASK EVENTS WITHIN THE WORK PACKAGE ARE COMPLETED
DIRECT PERCENT COMPLETE	VARIABLE	BUDGET EARNED BASED UPON OBJECTIVE ASSESSMENT OF COMPLETION STATUS OF TANGIBLE ACTIVITIES WITHIN THE WORK PACKAGE
DIRECT INPUT	VARIABLE	BCWP TAKEN BASED ON APPROVED SUBCONTRACTOR'S CPR OR C/SSR
UNITS COMPLETE	VARIABLE	BCWP TAKEN BASED ON QUANTITIES OF WORK INSTALLED
MILESTONE ACHIEVED	VARIABLE	BUDGET EARNED BASED UPON BUDGETED VALUE OF MILEDSTONES ACHIEVED
LEVEL OF EFFORT WORK PACKAGE	VARIABLE	BUDGET EARNED WITH PASSAGE OF TIME (BCWP ALWAYS EQUAL TO BCWS)
APPORTIONED EFFORT WORK PACKAGE	VARIABLE	BUDGET ON ALLOCATED BASIS IN DIRECT RELATIONSHIP TO BUDGET EARNED IN RELATED MEASURED CONTROL ACCOUNT/WORK PACKAGE

BCWP MUST BE EARNED USING THE SAME EARNED VALUE METHOD AS THAT USED TO DERIVE TIME-PHASED BCWS WHEN THE CONTROL ACCOUNT/WORK PACKAGE WAS INITALLY PLANNED ANS APPROVED.

Figure 8-2   Performance Measurement Methods

Report: CAP_LJ	Control Account Planning Sheet		Page : 4
Project: ANCDF	Description: PROJECT XYZ	Approval: Project Office	
Run Date: 08/03/03	Status Date: 07/31/03	Function Manager _____ _____ Control Account Manager _____ _____ _____ _____	

Control Account: 14313	Description: CHB CONCRETE	CA Manager: SMITH
Scheduled Start: 07/15/03	Scheduled Finish: 05/25/04	Status: Open

Work Package: 02	Description: CIVIL STRUCTURES		Scheduled Start: 07/15/03		Scheduled Finish: 05/25/04							
PMT: Milestone	Status: Open	Jul 03	Aug 03	Sep 03	Oct 03	Nov 03	Dec 03	Jan 04	Feb 04	Mar 04	Apr 04	May 04	At Complete
Cost Element FND/CIVIL MATERIALS	BCWS	16	28	39	55	71	110	83	50	50	28	20	550
	BCWP	50	0	0	0	0	0	0	0	0	0	0	
	ACWP	58	0	0	0	0	0	0	0	0	0	0	
	BCWS	14	23	36	49	66	102	76	45	45	25	19	500
	BCWP	45	0	0	0	0	0	0	0	0	0	0	
LABOR	ACWP	41	0	0	0	0	0	0	0	0	0	0	
Work Package Totals	BCWS	30	51	75	104	137	212	159	95	95	53	39	1050
	BCWP	95	0	0	0	0	0	0	0	0	0	0	
	ACWP	99	0	0	0	0	0	0	0	0	0	0	

Figure 8-3   Sample Control Account Plan

Use or disclosure of data contained on this sheet is subject to the restriction on the title page of this document.

Page 81

9. Material Planning and Control

Use or disclosure of data contained on this sheet is
subject to the restriction on the title page of this document.

Page 82

9.1 Overview

This section describes the requirements and methods used for planning, budgeting, controlling, and analyzing material. Material planning and control is concerned with the establishment of a material baseline for performance control which includes determining what materials are needed, in what quantity, at what price, and at what time. In addition, the material subsystem must provide for the timely recognition of price and usage variances. Figure 9-1 is an example of a material system flowchart, illustrating the basic material planning and control process.

9.2 Objectives

Material Planning and Control establishes guidelines for the timely control and analysis of materials. The primary objectives of the Material Planning and Control subsystem are:

☐ Plan material budgets (BCWS) for Work Packages and Planning Packages within Control Accounts in a manner consistent with the associated planned work.

☐ Accumulate/collect material cost (ACWP) in the same accounting month that performance (BCWP) is earned for that same material.

☐ Isolate, identify, and analyze price variances at point of commitment and update EAC at this point if appropriate.

☐ Isolate, identify, and analyze usage variances at various points in the material process as appropriate. Examples of material usage variances in the material process include purchase quantity, design quantity, scrap, shrinkage, etc.

9.3 Material Types

9.3.1 Material Categories

Materials are categorized as uniquely tagged items, bulk purchased commodities, and consumable materials and supplies.

Uniquely tagged items are specific items of equipment, assemblies, or engineered components that are unique to the contract and a Control Account. Uniquely tagged items are normally long lead and are assigned a "tag" number unique to the item. Project Controls normally assigns coding to identify the item to the Control Account using reference documents such as engineered drawings, the original estimate and the CWBS Dictionary. Each tagged item has a specified delivery date which is reflected in the schedule as a constraint to the start of the items installation activity. This item is budgeted (BCWS) based on either the scheduled receipt or issue date as appropriate for the item. The earned (BCWP) and actual (ACWP) costs are based on the same point in time, i.e., receipt or issue.

Bulk material commodities are materials procured based on Bill of Material (BOM) quantities. These materials are inventoried by type, i.e., 6" carbon steel pipe, 2" conduit, etc. Materials bought in bulk are purchased on a contract-wide basis and issued from the warehouse based on a warehouse issue ticket that specifies the Control Account number where the material will be used.

Consumable materials and supplies are low dollar items that are consumed in support of Work Package completion. Electrical tape, gloves, goggles, ropes and welding rods are examples of consumable supplies. Budget resources are scheduled based on historical experience and earned accordingly and, as such, are treated as an Apportioned Effort activity.

9.4 Material Planning and Control Process

9.4.1 Material Estimating

Material estimating is normally performed during the proposal cycle based on engineering drawings and equipment specifications. Takeoffs are made to arrive at bulk commodity quantities. Tagged items are grouped by like item, i.e., pumps, motors, etc. Actual vendor quotations are obtained on tagged equipment and estimated commodity material prices are applied to bulk quantities. Consumable supplies are estimated according to historical averages.

Use or disclosure of data contained on this sheet is subject to the restriction on the title page of this document.

Page 83

Estimated quantities are forwarded to the Proposal Team/Project Office for final review. Decisions are made at this time whether to include allowances for shrinkage, scrap, quantity changes, etc., and if so, how much. Once reviewed and adjusted as required, the quantity estimate is approved and becomes the basis for the contract estimate.

9.4.2   Material Budgets

Material budgets are based on the approved estimate plus or minus any changes that occurred during the contract negotiation and award process. The Project Office issues preliminary PWAs, which include material budgets, to Functional Managers. The Functional Manager issues preliminary CAWAs, which also include material budgets, to CAMs. The CAM divides the Control Account material budget into material Work Packages and develops the CAP. Once Control Account budgets are broken into Work Packages, a recap is made to ensure balance to the approved Control Account budget.

9.4.3   Performance Measurement

9.4.3.1   Tagged Items - BCWS Development

The CAM, using schedule delivery or issue dates, plans the budgets in the appropriate accounting period in the CAP. The resulting BCWS is loaded to the LGM INTERNATIONAL EVMS database.

9.4.3.2   Bulk Commodities - BCWS Development

These material budgets are reviewed by material type, i.e., carbon steel pipe, cable, conduit, valves, spares, repair parts, etc. The CAM, working with the Project Controls Scheduling Section, develops the material plan as a function of when the installation is scheduled to occur. Material BCWS is then planned based on the necessary issue (or receipt or installation, as appropriate) points required to support installation.

9.4.3.3   Consumable Supplies - BCWS Development

This type of budgeted cost is a function of the personnel intensity scheduled. As such, this is usually treated as an Apportioned Effort and BCWS is determined as a percent of the associated budgeted direct labor cost during each reporting period.

9.4.3.4   Tagged Items - BCWP and ACWP

As items are received, a Material Receiving Report (MRR) is prepared which notifies the CAM that an item has been received. At the point of receipt or issuance, as appropriate, the budgeted cost of the tagged item is earned (BCWP) and the actual cost is taken (ACWP).

9.4.3.5   Bulk Material - BCWP and ACWP Development

BCWP and ACWP for Bulk Material are determined utilizing various methodologies depending upon the scope of work for the Work Package and/or the material type.

☐ Earning Material at Point of Issue: When taking earned value at the warehouse issue point, ACWP is determined by the quantity of materials issued times the average inventory unit price at the time of issue. BCWP will be determined by the CAM based on the budgeted value for the scope of work represented by the issue.

☐ Earning Material at Point of Installation: In this situation, ACWP is determined by multiplying the actual installed quantity by the average inventory unit price for that quantity. BCWP can be determined by use of the previous two methods except that BCWP is determined at point of installation.

☐ Earned Value at Point of Receipt/Acceptance: ACWP is determined by multiplying the actual quantity of material received by the actual unit price. Again, BCWP is determined by the two previous methods except that BCWP is determined at point of receipt.

9.4.3.6   Consumables - BCWP and ACWP

Consumable materials are measured as Apportioned Effort in that they are related in direct proportion to a discrete labor. Consumable supply budgets are time-phased using a predetermined percentage of the associated discrete labor Work Package(s) and/or Control Account(s) BCWS. BCWP is earned at the same predetermined percentage of the associated Work Package(s) and/or Control Account(s) BCWP. ACWP is determined by the actual costs for the consumable supplies issued from inventory.

9.4.3.7   Usage Variance (UV)

A usage variance is generated when actual quantity varies from earned quantity and is calculated thus:

UV   = (Earned Quantity - Actual Quantity) X Budgeted Unit Price

Using pipe installation as an example, BCWS for the period was 18 linear feet (LF). At the time of installation, the design indicated a requirement of 19 LF. Further, assume that pipe is bought in 20 LF lengths, thus 20 LF is issued for the Work Package even though only 19 LF is required. In this example, BCWP is 18 LF and ACWP is 20 LF creating a usage variance of -2 LF: one foot due to increased requirement and one foot due to excess material (i.e., scrap). Assuming a budgeted unit price of $2/LF, a usage variance of $-4 exists.

9.4.3.8   Price Variance (PV)

Using the same example as with a usage variance, assume the 18 LF of pipe was budgeted at $2 per foot and the 20 LF was purchased for $3 per foot. A price variance is calculated thus:

PV = (Earned Price - Actual Price) X Actual Quantity

In this example:

$$PV = \frac{(\$2-\$3)}{LF} \times 20\ LF = \$-20$$

9.4.4   Budget Changes

Bills of Material (BOM) are generated based on drawing takeoffs and form the basis of the procurement cycle. As material requirements are identified, they are entered on a Purchase Requisition (PR) which when approved by the CAM and Functional Manager is forwarded to Project Controls for assignment of the Control Account code and represents the authorization document and material requisition for the procurement of materials.

If the budget is authorized, the revised budget is entered into LGM INTERNATIONAL EVMS and a revised CAWA is issued to the CAM. If the budget is not authorized but the material requirements are approved, the CAM must revise the EAC for that Control Account.

9.4.5   Material Receipt and Usage Analysis

As Purchase Orders (PO) are placed, purchased quantities are recorded in the materials subsystem and a copy of the executed PO is forwarded to the CAM, as appropriate, and to Project Finance/Accounting for payment process.

When materials are received, each delivery is checked and any purchase order material overages, shortages, or damages (OS&D) are noted. A Material Receiving Report (MRR) is generated for each material delivery by the warehouse. Received material is entered into the materials subsystem using the MRR. OS&Ds are noted and forwarded to the Project Procurement/Materials Manager for follow-up. The original MRR is filed in the warehouse and a copy forwarded to the CAM, as appropriate, and Project Finance/Accounting for use in the payment process.

9.5   Physical Accountability

Materials are drawn from the warehouse via a Warehouse Requisition (WR) approved by the CAM. All withdrawals are entered into the Material Subsystem to maintain material inventory balances and thus prevent intermingling of inventories, (i.e., separate inventories for a construction effort versus an operations effort). The Materials Subsystem produces reports that show, by commodity, materials issued by Control Account/Work Package and total balance in inventory.

9.6 Material System Flowchart

The material subsystem process flow, illustrated in Figure 9-1, is described below.

A. Upon contract award, the Project Office issues a work authorization to Procurement/Materials and the CAM. This authorizes the start of identification and procurement of required materials and initiates the planning and budgeting process.

B. Procurement/Materials develop a Materials Plan consisting of identification and categorization of tagged items and bulk materials. Responsibilities for spending limits are set; max/min requirements are established for bulk material to be kept in warehouse inventory, and the control spending limits for warehouse issues are set.

C. After development, the material plan is forwarded to the Project Office for review and approval. This is an interactive process resulting in the project material plan. This plan is returned to Procurement/Materials.

D. The CAM develops the CAP for material requirements and determines BCWS. Regardless of whether the material is long-lead (early buy) or bought based on requirements determined during the CAP process, BCWS is determined based on the schedule for receipt, issue or installation as appropriate as planned by the construction/operation site CAM. BCWP and ACWP are determined on the same basis.

E. Utilizing the CAP, the CAM defines the bulk material requirements and forwards them to Procurement/Materials.

F. The Functional Manager defines the tagged item requirements for the work authorized by the Project Office.

G. These requirements are noted on a material request and forwarded to Procurement/Materials.

H. Procurement/Materials collect and consolidate all tagged item requirements and bulk material requirements.

I. Based on the consolidated material requirements, Procurement issues Requests for Quotations (RFQ) to vendors.

J. Vendors receive RFQs and prepare the bids.

K. Priced bids are submitted to LGM INTERNATIONAL Procurement.

L. Procurement prepare bid tabulations for the materials being quoted.

M. The approved selection is forwarded to Procurement for a determination as to whether a purchase agreement exists or whether one needs to be established.

N. If the material is a bulk item, a master purchase agreement may be used for economic pricing or one is established as appropriate.

O. The Purchase Order is issued.

P. The vendor receives the Purchase Order and ships the appropriate material.

Q A copy of the purchase order is forwarded to the CAM. This is notification that materials have been ordered, what the materials are, when their delivery is scheduled and the committed value.

R. The CAM, after reviewing the purchased price will update the Control Account EAC if required.

S. Warehouse/Receiving receives the vendor shipment, inspects the materials and notes on the MRR, the items received, the quantity and the condition. If warranted, an OS&D report is filed for vendor follow-up.

T. The materials inventory is updated with the quantities received and the actual unit, lot prices and an updated MRR is generated.

U. A copy of the MRR is forwarded to Project Office Finance/Administration for payment backup.

V. The MRR is also forwarded to the CAM for notification that the materials are onsite, in inventory and available for issue.

W. A warehouse requisition is used to authorize the warehouse to release material from inventory. This requisition is approved by the CAM or designee.

Use or disclosure of data contained on this sheet is subject to the restriction on the title page of this document.

Page 86

X.   The warehouse will verify if the material is in inventory.

Y.   If the material is not in inventory, the warehouse verifies if the material has been ordered.

Z.   If the material is in inventory, it is issued to the CAM.

AA. If the material is not available or not available in the quantity required, the CAM is notified and any impacts to the plan are noted and alternatives plan developed.

BB. Required material not on order is requisitioned by the CAM. The requisition is forwarded to the Project Office for approval (Step M).

CC. Materials are issued to the CAM and the planned work is performed.

DD.   ACWP and BCWP are determined.

Use or disclosure of data contained on this sheet is
subject to the restriction on the title page of this document.

Page 87

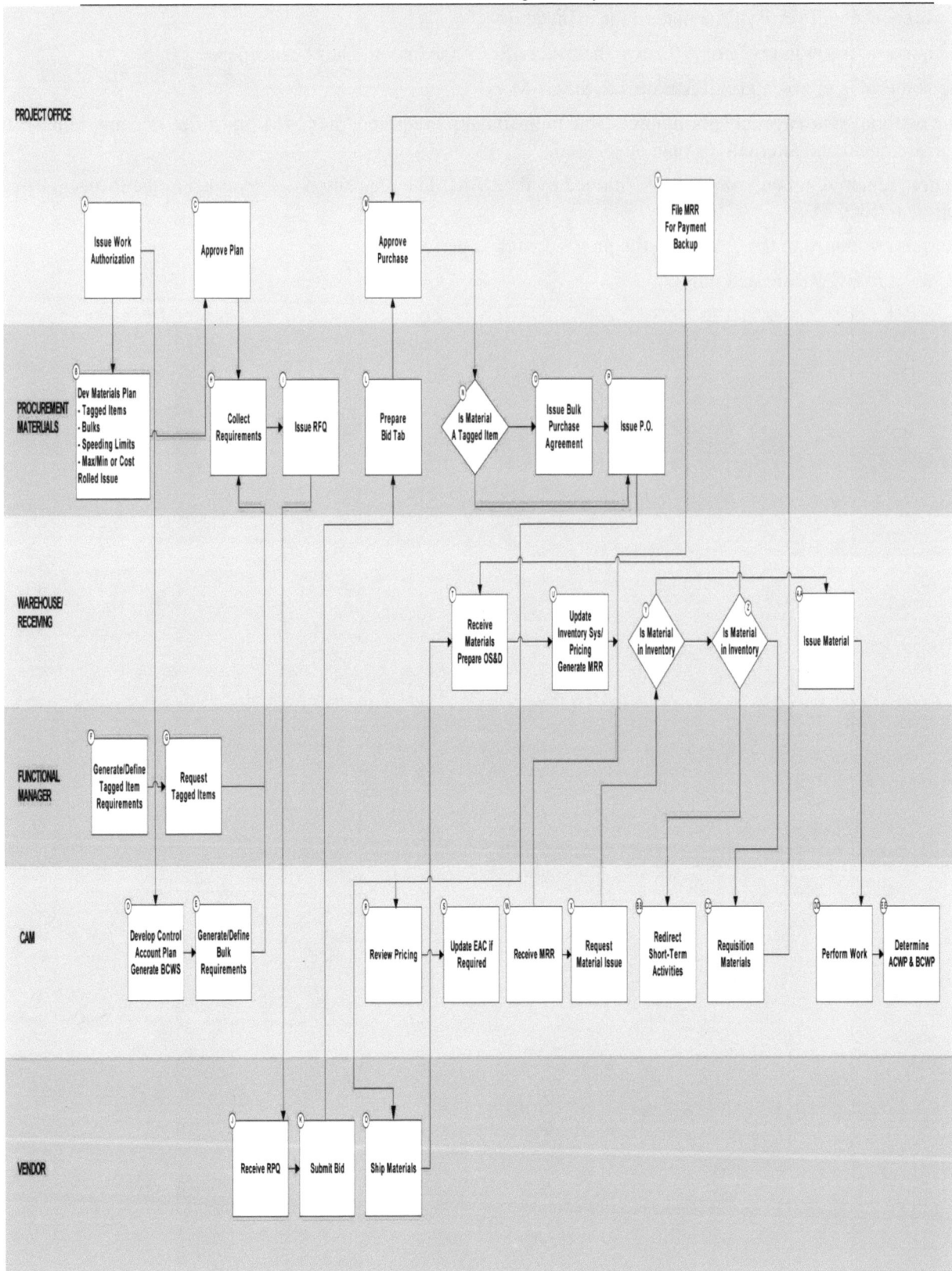

Figure 9-1   Material Subsystem Flowchart

Use or disclosure of data contained on this sheet is
subject to the restriction on the title page of this document.

Page 88

10. Subcontract Planning and Control

Use or disclosure of data contained on this sheet is
subject to the restriction on the title page of this document.

Page 89

10.1   Overview

The Procurement organization is responsible for acquiring materials, components, systems, services, construction contractors, operating equipment, and supplies from qualified sources, while achieving an optimum balance of schedule, price, quality, delivery, and performance consistent with LGM INTERNATIONAL policies and procedures, government regulations, and sound business practices. Major subcontracts require special treatment during both the pre-award and post-award procurement phases. The emphasis on oversight is normally due to the nature of the specialized services, equipment, construction and/or material being procured. The LGM INTERNATIONAL Subcontracting Procedures contain guidelines for subcontract management.

The purpose of this section is to explain the application of performance measurement requirements to selected subcontractors. Emphasis is placed on selected subcontracted areas of work that require the use of the following project management techniques, which promote more disciplined planning, managing, and control over subcontracted work. Major/critical subcontracts and non-major subcontracts include control and reporting provisions under a cost and schedule control system. Figure 10-1 lists the characteristics of major/critical, non-major, and minor subcontracts. Periodic status reports, reviews, and other appropriate management controls are used. These controls provide formal problem identification and reporting to ensure that potential problems are identified and acted on before project impact occurs. All major/critical subcontracts must maintain and use written operating procedures and criteria for planning, control costs and schedules, measure performance in terms of the value of work accomplished, and generate timely and reliable information for submission to the subcontract CAM.

10.2   Major Subcontract Definition

Major subcontract structures and systems are implemented on subcontracts that are:

a.   Selected by the customer and identified in the contract as a major/critical subcontract.

b.   Selected by the LGM INTERNATIONAL Project Manager during the make or buy determination (e.g., due to the criticality or uniqueness of the items to a contract end item and/or for a project for which its unique details can be bought only from the source of design and fabrication).

Flowdown of EVMS to the subcontractor will be by agreement between LGM INTERNATIONAL and the customer.

10.3   Major and Non-Major Subcontracts

10.3.1   Flowdown

As required by contract, major and non-major subcontracts are selected for application of the EVMS. This is based on criticality of the subcontract to the project, financial risk, technical achievability and schedule criticality. The level of detail in a subcontract CWBS is independent of the level of detail of the prime contract CWBS and is also independent of the level of the prime contract CWBS element into which the subcontract feeds. This means that if subcontracted work is large enough or complex enough to warrant an EVMS flowdown, then these subcontract work tasks may be broken down to the same extent as if the tasks were a prime contract. Care will be taken by LGM INTERNATIONAL to assure that the subcontractor has an appropriate CWBS in the subcontract. LGM INTERNATIONAL will contractually require the selected subcontractors to comply with the established reporting criteria. Adequate provisions for an EVMS demonstration or review, acceptance and surveillance of the subcontractor's systems shall be a subcontract requirement and carried out by LGM INTERNATIONAL or the customer.

10.3.2   Reviews

In the event that a subcontractor's system was previously approved by a qualified government agency, LGM INTERNATIONAL shall accept the approval as satisfying contractual EVMS requirements. For EVMS flowdown, assuming previous approval of the subcontractor's system, LGM INTERNATIONAL will conduct or cause the system to be subjected to an Integrated Baseline Review (IBR).

Use or disclosure of data contained on this sheet is
subject to the restriction on the title page of this document.

Page 90

If the subcontractor is utilizing an EVMS which has been previously accepted or is operating such systems under a current Memorandum of Understanding or Advance Agreement by the Government, the Contracting officer may waive all or part of the provisions concerning demonstration and review.

Should the subcontractor not have an approved system, LGM INTERNATIONAL, the Government and the subcontractor will mutually agree on either a Government or a LGM INTERNATIONAL conducted review. A LGM INTERNATIONAL conducted review will use the Government's EVMS documents as a guide; however, LGM INTERNATIONAL approval will be for the sole purpose of the affected contract. A review and acceptance of a subcontractor's system by LGM INTERNATIONAL does not constitute an approval by the Government.

Generally, LGM INTERNATIONAL will request Government surveillance assistance for a subcontractor's system only when:

1. LGM INTERNATIONAL is unable to accomplish the required surveillance because it would jeopardize the subcontractor's competitive position or proprietary data is involved.

2. There is a business relationship between LGM INTERNATIONAL and the subcontractor not conducive to independence and objectivity, as in the case of a parent-subsidiary or when LGM INTERNATIONAL and subcontracting roles of the companies are frequently reversed.

3. The subcontractor is sole source and the subcontract costs represent a substantial part of LGM INTERNATIONAL costs.

10.3.3   Surveillance of Subcontractors EVMS

When an EVMS is flowed down to a subcontractor's management system, surveillance becomes an integral part of the management and administration of a subcontract. The surveillance is to be carried out by LGM INTERNATIONAL, or by the Government, if agreed to, when requested by either LGM INTERNATIONAL or the subcontractor. Access to all pertinent records, documents, supporting information, instructions, and data shall be a subcontract requirement and be provided to LGM INTERNATIONAL and, upon request, to the Government Contracting Officer or duly authorized representative for the purpose of permitting LGM INTERNATIONAL and/or the Government to assure implementation and continuing application of the subcontractor EVMS-compliant management system requirements.

10.3.4   Major and Non-Major Subcontracts Reporting

Associated subcontract integrated cost/schedule status reporting requirements (e.g., CPR, C/SSR, Contract Funds Status Report [CFSR]) will be obtained monthly on major subcontracts other than firm fixed-price purchase subcontracts to properly assess performance during the life of the subcontract. The CPR is identified in the subcontract as a separate SDRL when EVMS is a flowdown requirement. When a subcontractor is required to comply with the EVMS and provide a CPR, subcontractor data will be provided to LGM INTERNATIONAL for performance measurement purposes. If a subcontractor is not required to comply with the criteria, LGM INTERNATIONAL may establish procedures which tie the planned and actual accomplishment of the subcontractor to valid indicators, such as the proposed payment schedule or completion of identified work segments. The C/SSR may be used for obtaining the data from a subcontractor when EVMS and the CPR are not required. When a C/SSR is a subcontract requirement, the C/SSR, a time phased ETC report, the Baseline (CPR Format 3), the CFSR, and schedules are included in the SDRL as requirements. Normally, the SDRL specifies that the CPR or C/SSR data will include subcontractor fee and the subcontractor will include, on a quarterly basis, a reconciliation of the CPR or C/SSR to the CFSR. The ETC report is usually required only when the total EAC or the time phasing of the ETC has been changed by the subcontractor.

If the subcontract effort is not contained within one lowest level LGM INTERNATIONAL CWBS element, the SDRL will specify that a separate Baseline (CPR Format 3) will be submitted for each affected CWBS element by the subcontractor to facilitate determining BCWS for each LGM INTERNATIONAL Control Account. In these cases, the SDRL will normally specify that a separate breakout of the data includes the allocation of MR, UB, and any other reporting elements which are common to the LGM INTERNATIONAL CWBS elements within which the subcontract effort is contained. If this is not required of the subcontractor, the CAM must accomplish the allocation to the various Control Accounts prior to the input of data into LGM INTERNATIONAL EVMS. An alternate, but not preferred,

Use or disclosure of data contained on this sheet is
subject to the restriction on the title page of this document.

Page 91

method is to include the subcontractor's MR and UB in LGM INTERNATIONAL MR and UB and separately identify these in the LGM INTERNATIONAL CPR or C/SSR Narrative Variance Analysis.

The extent of reporting will be determined by LGM INTERNATIONAL contract requirements, subcontract value and duration, criticality of the subcontract item to the prime contract, and past performance of a specific supplier on previous subcontracts. Electronic Data Interchange (EDI) for transmission of subcontractor's CPR, C/SSR, CFSR, etc. will be utilized where possible.

Those subcontracts with a budgeted value in excess of $1 million and which meet any one of the following criteria are also considered either a major or non-major subcontract:

☐ Other than firm-fixed price (FFP)

☐ Any subcontract (including FFP) with either progress or milestone payments and which has six months or more between the beginning of work and the first significant delivery

☐ Any subcontract (including FFP) which represents significant schedule or technical risk to the project, as identified by the Project Manager in coordination with the Customer

10.3.5 Non-Major and Minor Subcontract Reporting

Non-Major subcontracts are not considered for CPR or C/SSR reporting when they are firm fixed price or fixed price with escalation type subcontracts, or when their cost, schedule criticality, or design criticality do not warrant CPR or C/SSR reporting. Figure 10-2 presents cost guidelines for subcontractor reporting requirements. Control of these non-major and minor subcontracts will be tailored to the specific non-major or minor subcontract. The subcontract administrator will normally negotiate a performance report with the subcontractor to ensure that the Government and LGM INTERNATIONAL receive proper visibility into the subcontractor's performance. Figures 10-2 and 10-3 provide additional guidance on reporting requirements.

10.4 Subcontract Planning, Performance Measurement, and Actual Cost

Normally, a separate CAWA is prepared and issued for each subcontract. This allows for summarization of the budget and performance measurement data to a single level prior to summarizing it with other subcontracts to the total functional organizational level.

Major subcontracts are planned in separate Control Accounts (i.e., Control Accounts that include subcontracted effort only). If the subcontracted effort falls into more than one CWBS element, LGM INTERNATIONAL attempts to negotiate with the customer for reporting in only one element (i.e., the element with the majority of the effort). If this is not possible, Control Accounts are established for each CWBS element that contains effort. Normally, the same CAM (the subcontract coordinator) is responsible for all the Control Accounts for that subcontract.

All subcontract planning, performance measurement, and actual cost values represent LGM INTERNATIONAL's cost (i.e., they include the subcontractor's fee or profit). If the subcontractor submits reports at the cost level (i.e., excluding fee), LGM INTERNATIONAL will calculate fee and add it to the values provided by the subcontractor prior to entering it into LGM INTERNATIONAL EVMS.

Subcontracts are normally budgeted by LGM INTERNATIONAL at the value of the subcontract. When this is not the case, LGM INTERNATIONAL factors all data with the exception of ACWP, received from the subcontractor using the following formula prior to entering it into LGM INTERNATIONAL EVMS:

LGM INTERNATIONAL Budget at Completion (BAC) for Subcontractor
Subcontractor's Negotiated Price at Completion

10.4.1 Subcontract Planning: Budgeted Cost for Work Scheduled

For subcontracts that require cost performance reporting, BCWS is planned in accordance with the baseline plan provided by the subcontractor. If the subcontractor's report does not include fee or profit, the CAM adjusts the baseline plan received from the subcontractor to include the subcontract's target fee/profit. (If the plan includes this fee, no

Use or disclosure of data contained on this sheet is subject to the restriction on the title page of this document.

Page 92

adjustment is necessary). If more than one Control Account per subcontract is required or a C/SSR is required, the baseline plan reports must be requested in the contract data requirements list. If performance reports cannot be received in time to incorporate them into LGM INTERNATIONAL's current month data, then they will be lagged by one month. When this is the case, BCWS is also lagged by one month. Normally, only one Work Package is necessary for each subcontract Control Account.

The BCWS for Control Accounts for subcontracts that do not require performance reporting is developed by the responsible CAM. It may be based on the subcontractor's billing plan, a major milestone or delivery schedule, or other available information (see Figures 10-1 and 10-3). (Again, target fee/profit is included.) The CAM must, of course, be able to explain the basis for BCWS and must measure performance (BCWP) in the same manner.

The following types of subcontracts do not normally require performance reporting from the subcontractor:

a. Time and material subcontracts for the purchase of a fixed level of resources to be expended over a set period of time. For this type of contract, the CAM develops BCWS based on the subcontractor's expenditure plan and the earned value technique is LOE.

b. For firm-fixed-price subcontracts without progress payments, the subcontractor's schedule, payment plan and/or scheduled milestone/delivery dates should be submitted.

c. For subcontracts with progress payments where the subcontractor is unwilling to submit a performance report or the proposed cost of such a report is not considered practical or reasonable, the CAM will plan BCWS based on the best knowledge of the subcontract work to be performed and by using other existing company and subcontractor information, such as schedules, technical reports, etc.

10.4.2 Subcontract Performance Measurement: Budgeted Cost for Work Performed

BCWP for subcontracts which provide CPRs or C/SSRs is derived directly from those reports. Again, the information is adjusted to include the subcontractor's target fee/profit if the performance report has been provided at the cost level. It is entered into the LGM INTERNATIONAL EVMS.

The BCWP for subcontracts that do not require performance reporting is developed by the responsible CAM, on the same basis that BCWS was established. Typically, BCWP is the subcontractor's milestone accomplishments claimed and approved during the month.

Time and material subcontracts earn BCWP using the LOE method. Firm-fixed-price subcontracts that have no progress payment earn BCWP based on actual deliveries and/or milestone accomplishment.

10.4.3 Subcontract Actual Cost of Work Performed

The CAM uses the CPR or C/SSR reports to derive ACWP. As with BCWS and BCWP, the CAM adjusts the subcontractor's submitted data to sell price as necessary. Factoring is not accomplished against actuals. The CAM reconciles the subcontractor reported ACWP values with the subcontractor's billings or actuals in the formal accounting records by comparing the CPR or C/SSR to the subcontractor's CFSR.

For FFP subcontracts with CPR or C/SSR requirements, if the subcontractor's ACWP is not equal to the BCWP, the CAM adjusts ACWP to be equal to BCWP before submitting the data to the Project Office for input into LGM INTERNATIONAL EVMS.

For subcontracts without CPR or C/SSR reporting requirements, the CAM must input the ACWP into the LGM INTERNATIONAL EVMS in the same accounting month that BCWP is earned. For FFP subcontracts, ACWP is set equal to BCWP. For other than FFP contracts without C/SSR reporting, the ACWP is set equal to the invoice price of milestone billing, progress payments, or their actuals submitted by agreement. If holdbacks of bills paid are part of the subcontract, the ACWP should include the number of holdbacks. For time and material LOE subcontracts, ACWP is set to the invoice values for the month.

The above cases for ACWP are considered estimated ACWP and are directly input into LGM INTERNATIONAL EVMS. Actuals from the Accounting System are not used in these cases, but adjustments to actuals in LGM INTERNATIONAL EVMS may be necessary periodically to agree with the actuals in the Accounting System.

10.5   Subcontractor Variance Analysis and Estimates at Completion

A VAR is generated for all subcontract Control Accounts which exceed the established thresholds. The CAM, with the support of Project Controls, completes the VAR following the normal review and approval cycle. When a CPR or C/SSR is submitted by the subcontractor, the CAM reviews the variance analysis included in it and uses it as an input in preparing the Control Account VAR.

As part of monthly analysis review, the CAM reviews the subcontractor-reported EAC. Based on this analysis, the CAM either submits the value as reported or reflects his or her independent EAC value. The subcontractor's time-phased ETC is used as a basis for the CAM's time-phased ETC. If the CAM's ETC differs from the subcontractor's ETC, the responsible CAM must support and justify which value is used on the VAR. Any EAC changes to the Control Account for subcontracted effort must be approved by the Project Manager in the same manner as internal Control Account EAC changes.

For FFP subcontracts with CPR or C/SSR reporting, the CAM must ensure that the EAC submitted to the Project Office for input into LGM INTERNATIONAL EVMS is always equal to the BAC (i.e., the EAC for an FFP subcontract should always indicate an on-target forecast unless the Control Account budget does not equal the negotiated value of the subcontract). In these cases, factoring does not impact the EAC. However, if the subcontractor is indicating an overrun on the FFP subcontract (subcontractor paying for the overrun) or an under run (subcontractor increased effective profit); this information should be included by the CAM in the narrative portion of the VAR.

SIZE OF SUBCONTRACT	TASK	COMPLEXITY	VALUE/TYPE	REPORTING REQUIREMENTS
Major/Critical Subcontracts	• New R&D • Major redesign • Critical interface to prime • Unique subsystem	• System definition not firm • Changes anticipated throughout • Mission related requirement • Multi-year development • Customer requirements change during development • Complex integration and testing	• More than $73M RDT&E • More than $315M Prod. or O&S • CPFF, CPIF/AF, FPI/AF • PDR, CDR, FDR milestones • Bill and fee share arrangements	• Full CPR or C/SSR Reports • CFSR • Quarterly evaluation for fee • Full technical and milestone accomplishments • Time-phased ETC
Non-Major Subcontracts	• Modifications to developed equipment • Limited software development • State-of-the-art product line systems • Interface to prime subsystem	• Mission related requirements • Leading edge of technology • Requirements may exceed current performance • Less than multi-year development • Acceptance testing required	• Less than $73M RDT&E • Less than $315M Prod. or O&S • FPI or FFP • PDR, CDR, FDR milestones • Some variety of progress payments	• C/SSR and CFSR Reports • No-fee evaluations • Milestone statusing • Time-phased baseline • Time-phased ETC
Minor subcontracts	• Very minor modifications • No critical interface	• Requirements firm • Not mission related • Less than 6-month delivery • Reliance on vendor representation as to performance • No complex acceptance tests	• Under $6.3M • FFP • No design milestone • Payment on delivery	• No financial reports • Progress reports (if required) • No tracking of budgets • Milestone statusing
The distinction between non-major and major/critical is dependent on the task, complexity, size, and type of subcontract. Discussion of minor subcontract is for information only.				

Figure 10-1   Minor, Non-Major, and Major/Critical Subcontract Definition

Use or disclosure of data contained on this sheet is subject to the restriction on the title page of this document.

Page 95

	DEVELOPMENT			PRODUCTION		
	More than	$6.3M to $73M	Under $6.3M	More than $315M	$6.3M to $315M	Under $6.3M
*CPR	M	R	N/R	M	N/R	N/R
C/SSR	R	R	N/R	R	R	N/R

Legend: M   Mandatory

              R   Recommended

            N/R   Not recommended

*CPR is required when the subcontractor has EVMS as a requirement.

Column headings refer to target price (FPI or CPIF) or estimated price.

Size of contract (for purpose of this chart) includes cost of options, anticipated modifications, or other changes.

Figure 10-2   Subcontract Reporting Requirements

Use or disclosure of data contained on this sheet is
subject to the restriction on the title page of this document.

Page 96

TECHNICAL AND/OR SCHEDULE RISK FACTORS			
	HIGH	**MEDIUM**	**LOW**
H I G H	CPR with Approved EVMS	CPR Required With or Without EVMS	C/SSR Required

M E D I U M	CPR Required With or Without EVMS	C/SSR Required	• Spend Billing Plan
			• Major Milestone Chart
			• Monthly Status Report
			  - Technical
			  - Schedule
			  - Cost

L O W	C/SSR Required	• Spend Billing Plan
		• Major Milestone Chart
		• Monthly Status Report
		  - Technical
		  - Schedule
		  - Cost
			Monthly Letter Report

(Left axis: **COST FACTOR**)

- Consider Risk Factor
- Consider Cost Factor
- Consider Schedule Constraints and Delivery Dates
- Consider Engineering Design Effort
- Consider History
- Consider Main Product Line
- Consider Inspection Process
- Consider Alternate Supply Sources

Figure 10-3   Technical and/or Schedule Risk Factors

11. Cost Accounting

11.1 Overview

The cost accounting system has been developed using a standardized coding framework. This framework identifies the essential information required to correctly collect cost charges for specific items of the work process. This coding structure is used on all accounting transactions to define the levels and categories required to apply costs to each Control Account.

LGM INTERNATIONAL utilizes various interactive computerized subsystems such as payroll, material control, and accounting to record and process actual costs. The uniform coding structure allows for the actuals to be segregated into such categories as labor hours and labor, material, subcontracts, ODCs, Overhead/G&A dollars for reporting and reconciliation purposes.

Costs are accumulated as either direct or indirect. Direct costs are labor, material, subcontracts and ODCs identified to benefiting Control Accounts as incurred or costs applied directly to Control Accounts through a distribution system. Indirect costs are classified as overhead or G&A costs which benefit two or more projects and cannot be consistently or economically identified directly to project effort are chargeable to expense accounts and subsequently applied to programs based on specific direct costs as a distribution base. Overhead/G&A costs are discussed more fully in Section 12.

11.2 Cost Accounting Policy

Financial information will be properly documented, approved, recorded, and processed in accordance with written corporate policies and procedures conforming to the requirements of Generally Accepted Accounting Principles. LGM INTERNATIONAL corporate policies and procedures satisfy the requirements in the Cost Accounting Standards (CASs) and the cost principles and procedures in Part 31 of the Federal Acquisition Regulation (FAR) 31. These systems are audited by the Defense Contract Audit Agency (DCAA) and approved by the cognizant Corporate Administrative Contracting Officer at the Defense Contract Management Agency (DCMA). Corporate policies and procedures also accommodate specifications set forth by the following organizations:

- Accounting Principles Board of the AICPA
- Accounting Research Bulletins
- CAS Board
- Financial Accounting Standards Board
- Internal Revenue Service
- National Association of Accountants
- Securities and Exchange Commission

LGM INTERNATIONAL Administration and Finance oversees corporate compliance with applicable Federal Regulations and maintains control over written corporate policies and procedures. This group monitors and maintains the system of internal accounting controls to ensure reliability and adequacy of its records (books) and accurate, timely control and reconciliation of all financial transactions.

LGM INTERNATIONAL utilizes a calendar Fiscal Year divided into 12 fiscal months. The fiscal month ending calendar is established prior to each calendar Fiscal Year, and consists of four or five week months. The accounting cycle is closed at the end of each fiscal month to summarize all financial information for project and corporate ledger reports. Standard accounting close is typically the last Saturday of the month. The monthly closeout consists of five cycles. Each closeout cycle represents deadlines for project and corporate accounting ledger reporting. Ledger reports operate at a contract level and detail current month transactions and provide year-to-date and inception-to-date totals by cost type. All ledger data are adequately supported by auditable documentation which is referenced to provide a traceable audit trail. Figure 11-1 illustrates the accounting process.

Use or disclosure of data contained on this sheet is subject to the restriction on the title page of this document.

Page 99

11.3 Cost Accounting Summarization

The cost accounting system is structured to provide summarization of direct costs from the Control Account into the CWBS elements without allocation of a single Control Account element to two or more CWBS elements. It is also structured to provide summarization of direct costs from the Control Account into the OBS without allocation of a single Control Account to two or more OBS elements. This capability permits the entry of the actual direct cost data into the system at the Control Account level and to accumulate the data for reporting at any level of the CWBS or OBS where budgets have been established.

The integrity of the accounting information is assured by the following features:

A. Correlation of cost/schedule status information is ensured by reporting cost and schedule performance data at the end of each accounting month in accordance with the fiscal month ending calendar.

B. The baseline control system prohibits unauthorized changes to tasks, definition, budget, and schedule within a given Control Account so that earned value will relate directly to the approved authorized scope of work.

C. The accounting code structure provides the link between BCWS, BCWP, BAC with ACWP, and EAC for a given Control Account. For example, this link enables accurate cost accumulation by element of direct cost and assignment to Control Accounts in a manner consistent with budgets.

D. Apportioned Effort cost is charged directly to Apportioned Effort Control Accounts or Work Packages; there is no allocation of these costs. ACWP used for variance analysis is reconcilable with data from the cost accounting system. All direct costs charged to a contract are auditable to accounting records.

E. The accounting code structure assigned to a given Work Package directly links it to the correct Control Account. All codes are validated before final posting in the cost accounting system. Any incorrect or incomplete codes will not be accepted and must be corrected before final posting in the system.

F. The monthly performance report provides summarized performance measurement data from the Control Account level to the appropriate reporting levels of the CWBS and OBS.

G. Control Accounts are opened when work on the first Work Package is scheduled to start, or actually starts if authorized ahead of schedule. Control Accounts are closed when the last Work Package is completed.

H. Project Controls, upon receipt of notification from the CAM, updates the LGM INTERNATIONAL EVMS records to close out the account number to all future charges.

I. Valid charges to the completed Control Account arriving after the closure of the account number are recognized by Project Controls and returned to the CAM for verification. The CAM verifies the charges and returns the verification papers to Project Controls. Project Controls reopens the closed account number and enters the charges into the Control Account.

J. The CAM has five (5) work days to verify the late charges to the Control Account. If the CAM fails to verify the late charges, Project Controls, by default, will enter the late charges into the closed Control Account.

K. If a late charge is identified by the CAM as invalid, it is so noted and the verification paper is returned to Project Controls. Project Controls returns the late charges to the source responsible for the charge correction.

L. When disputes arise pertaining to late charges as to validity between the CAM and the source responsible, the Project Office will make the final determination of the validity of the charge

M. Retroactive changes to direct and Overhead/G&A costs are prohibited except for the correction of errors and routine accounting adjustments. Direct and Overhead/G&A cost adjustments are made according to the CAS Disclosure Statement and Standard Accounting Procedures and principles which are acceptable by DCAA.

11.4 Accounting Code Structure

The corporate coding structure is divided into three major sections, as depicted in Figure 11-2.

Use or disclosure of data contained on this sheet is subject to the restriction on the title page of this document.

Page 100

11.4.1 Organization Codes

These codes identify Company, Accounting Center, and Responsibility Center information. The Company code identifies a legal entity within the consolidated group of LGM INTERNATIONAL subsidiaries and affiliates. The Accounting Center code denotes a location where a balanced set of accounting records and accompanying support documents are maintained. The Responsibility Center code specifies the product line sponsoring and having ultimate responsibility for the project.

11.4.2 Accounting Codes

These codes identify Job Number (or contract), Billing Type, and General Ledger categories. The Job Number code identifies a specific endeavor for which Control Accounts are maintained and reported. This code is used to consolidate all accounting transactions performed under a specific contractual agreement. The Billing Type code identifies the billing category for a particular transaction and provides for segregation of job revenue and cost information. The General Ledger codes identify ledger accounts that allow cost to be segregated by element of cost.

11.4.3 Project Codes

These codes identify project specific Control Accounts and Work Packages to allow for actual costs to be collected at the same levels and categories that the work was budgeted.

This standardized coding structure allows for summarization of all recorded accounting transactions at any level. In certain cases, accounting codes may be consolidated to summarize actual costs to CWBS and OBS elements without allocation. By combining the General Ledger and project codes, any level of comparison can be made for ACWP with any budgeted information. Predetermining in LGM INTERNATIONAL EVMS of quantity units of measure enables the cost accounting system to determine unit costs, equivalent unit costs, or lot costs in terms of all budgeted element of cost.

11.5 Recording Cost

ACWP is cost which, by its nature, is either immediately identified or otherwise applied to the contract. As direct resources are applied toward the accomplishment of authorized work within a given Control Account, actual costs are recorded in the accounting system identified by the Control Account number. By assignment of specific CWBS elements as recurring and non-recurring, LGM INTERNATIONAL EVMS has the ability to report recurring and non-recurring costs. Control Account data are recorded by element of cost in terms of monthly and cumulative (inception-to date) expenditures in a manner consistent with budgets using recognized acceptable costing techniques and controlled by the general books of account (General Ledger System).

11.5.1 Direct Labor

Labor hour expenditures for project activities are recorded on timesheets. These timesheets are the initiating source for actual labor hour information and subsequent payroll and labor reports.

Labor hours for construction labor are recorded for each member of the construction crew by the crew foreman on a daily timesheet in accordance with established timekeeping practices. These timesheets include the appropriate codes for the work performed, a description of the work, the number of hours expended on each work assignment and the total labor hours. Each of these timesheets is reconciled with each individual's attendance records by the timekeeper. Other direct labor personnel such as Engineering and Home Office are responsible for completing their own timesheets which are then reviewed by each employee's supervisor for approval in accordance with LGM INTERNATIONAL's timekeeping system. The approved labor hour data are then processed through the LGM INTERNATIONAL payroll systems which are posted on a weekly basis. Each employee's labor rate is used to determine the actual labor cost charged to each specific Control Account. When the payroll system passes the labor information to the Performance Measurement Subsystem, each charge is validated against the Control Account coding structure to ensure proper coding. The direct labor costs are accumulated in the cost accounting system and reported

Use or disclosure of data contained on this sheet is subject to the restriction on the title page of this document.

Page 101

internally each week. These charges can be summarized by CWBS and OBS elements and then reported as ACWP during the monthly reporting cycle.

11.5.2   Material and Subcontract Costs

LGM INTERNATIONAL EVMS monitors materials and subcontracts by recording each purchasing document in the material system. Each commitment is coded using the general books of account and project coding structure. The general books of account record actual payments to vendors/suppliers based on approved Material Receiving Reports (for material/supplies physically delivered) and on approved Inspection Reports for vendors/subcontractors where progress payments are made. ACWP is recorded in the Performance Measurement subsystem on an estimated actual cost basis. Due to the cash basis for material accounting in the general books of account, there is a time differential between ACWP in the Performance Measurement subsystem and the costs recorded in the general books of account. Reconciliation of material cost is accomplished by a comparison of material ACWP in the Performance Measurement subsystem, Material Inventory value in the material subsystem, and the material open commitment value also carried in the material subsystem. Open commitments will be relieved as actual vouchers are recorded in the general books of account rounding out the reconciliation process.

Final reconciliation can be performed at project completion through a comparison of ACWP, residual inventory value and cash payments recorded in the general books of account.

11.5.3   Other Direct Costs

Principle ODC items such as travel, data processing, reproduction, etc., are collected in the financial accounting system and reported monthly. After closeout, a detail cost ledger that collects these ODC items by general ledger code is produced. Project Controls uses this monthly detail cost ledger to assign ODC cost into the appropriate Control Account. After validating and reconciling the codes, these ODC actual costs are posted into the cost accounting system for reporting purposes.

11.5.4   Indirect Costs

Indirect costs are accumulated into cost pools and applied to programs based on the cost allocation requirements described in LGM INTERNATIONAL's CASB Disclosure Statement. Indirect rates are applied to each project at the rates negotiated with DCMA in accordance with FAR 42.17, Forward Pricing Rate Agreements (FPRA). Indirect costs are budgeted at the Control Account level for tracking purposes (if directed by the customer) or, preferably at the total contract level. In either case, Project Controls (not the CAM) is responsible for the analysis at the total contract level.

LGM INTERNATIONAL's Disclosure Statement explains in detail the process concerning the collection and reporting of direct and indirect costs. DCAA reviews the Disclosure Statement and makes recommendations to the CACO who is responsible for CAS administration and approval of the Disclosure Statement.

The DCAA reviews the support documents (time cards, invoices, journal entries, purchase orders and vendor payments, etc.) and the management approval system, on an ongoing basis, to ensure that system integrity is maintained in conformance with government regulations. Through these types of reviews the accounting system provides a basis for auditing records of direct costs chargeable to the contract.

11.6   Accounting Adjustments

Retroactive adjustments to accounting data are prohibited except for the correction of errors and the normal direct or indirect rate adjustments required to maintain accuracy. Annual indirect rates and claimed direct costs are monitored monthly against actual recorded cost to maintain an accurate projection of the total costs. Periodically, LGM INTERNATIONAL analyzes its indirect cost rates and if the analysis discloses that the rate is under or overstated to the point that a large year-end adjustment would be required, we negotiate an update to the FPRA with DCMA and costs recorded or charged are adjusted for the entire company fiscal year to date. All accounting adjustments are available for audit and review by the DCAA.

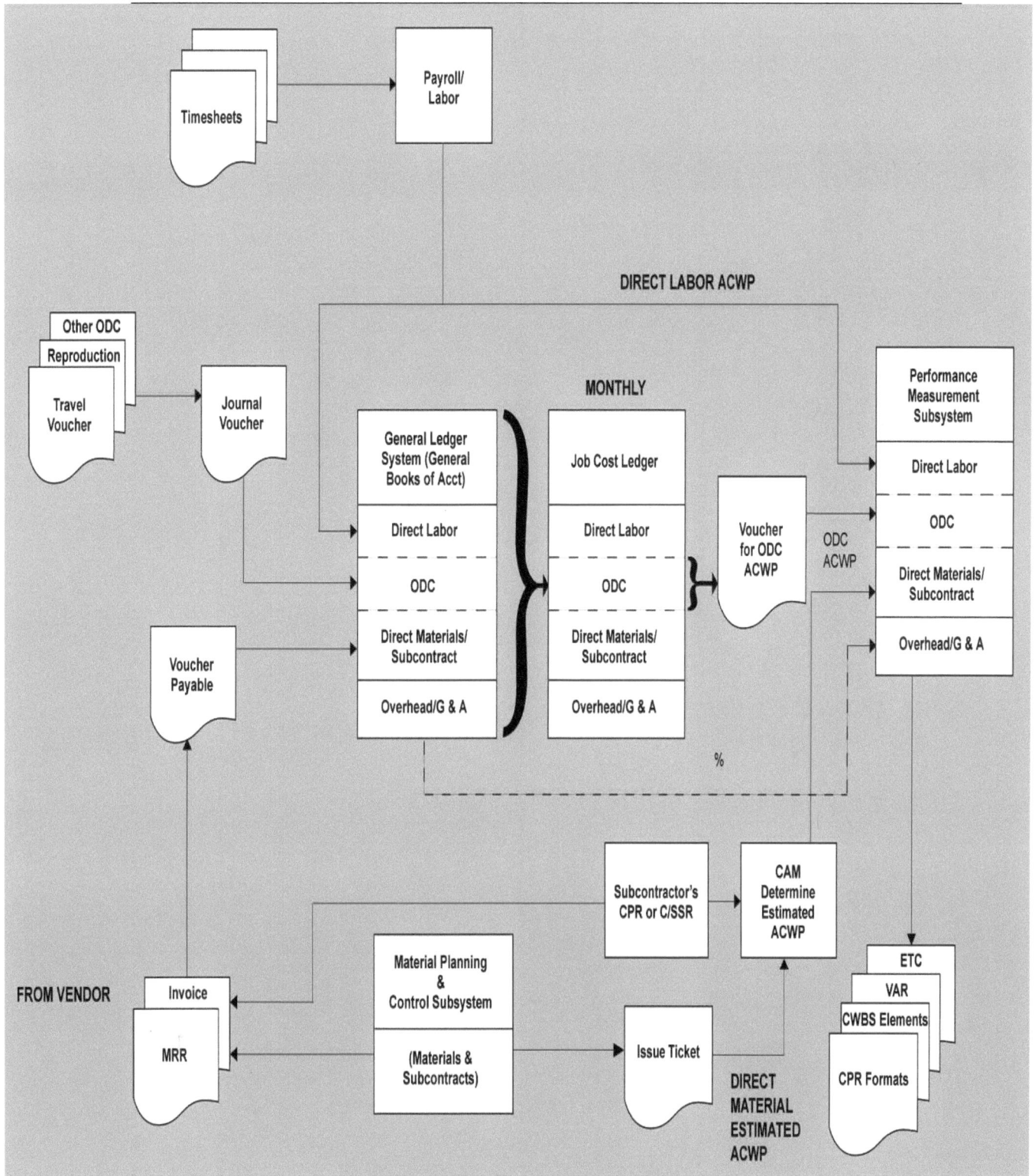

Figure 11-1   Accounting Process

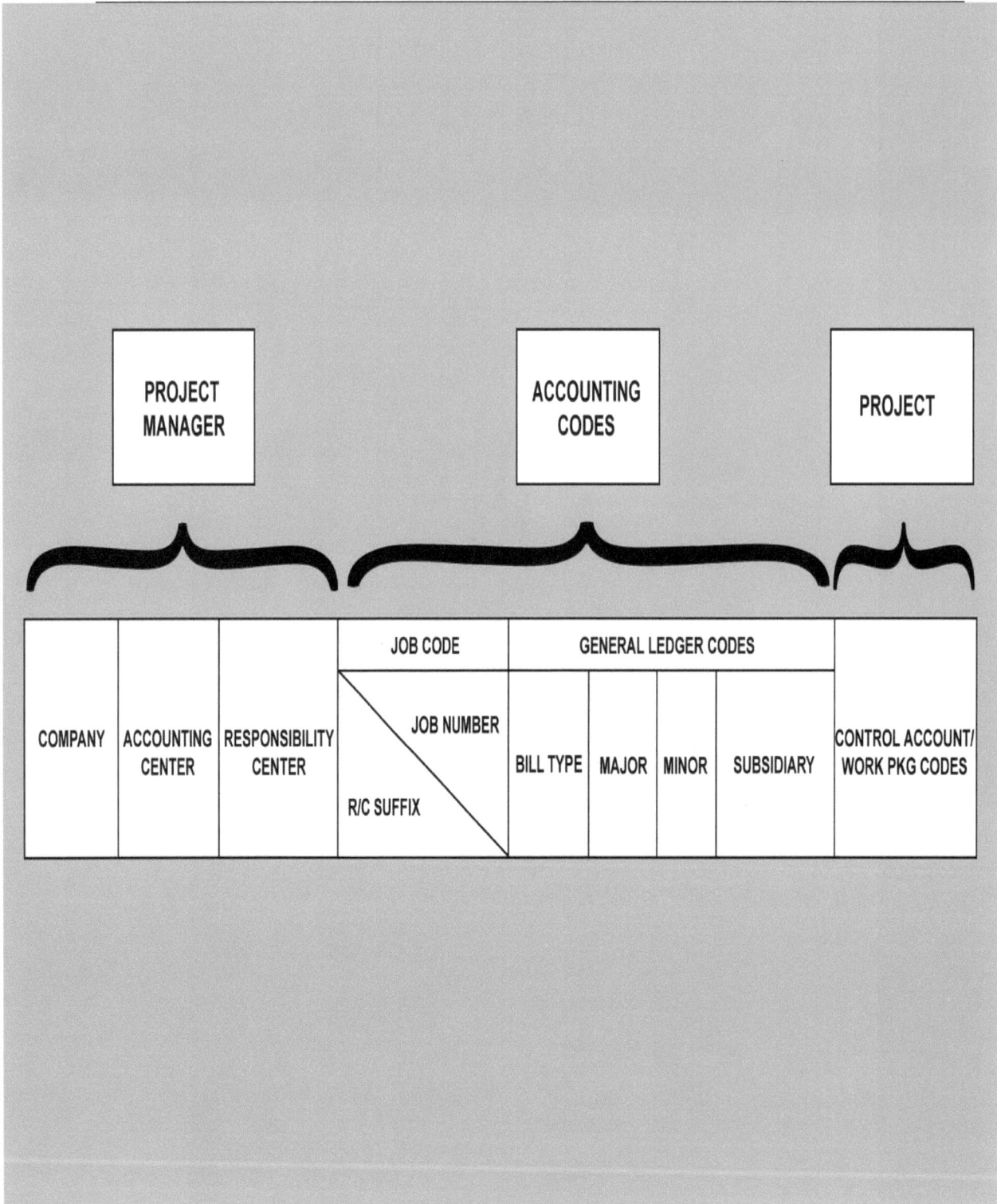

Figure 11-2   Corporate Coding Structure

Use or disclosure of data contained on this sheet is
subject to the restriction on the title page of this document.

Page 104

12 Overhead Cost Management

12.1 Overview

Indirect costs are those that, because of its benefit or incurrence in support of many common or joint objectives, are not readily measurable to the extent that the exact amount of the cost pertaining to a specific cost objective can be determined. Indirect costs are those necessary to maintain the organizational elements which support multiple contract efforts.

LGM INTERNATIONAL is an organization consisting of subsidiary corporations and affiliates. In accordance with our disclosed practices, LGM INTERNATIONAL has the following major areas of indirect cost:

Indirect Cost Pool	Cost Allocation Base
Overhead	Total Project Cost
Procurement Service Center – CONUS	PSC domestic direct labor
Procurement Service Center – OCONUS	PSC OAS/SEII direct labor
General and Administrative	Total Cost Input

These pools and bases are described in the following LGM INTERNATIONAL Disclosure Statements:

1. LGM INTERNATIONAL – Parts I through VII – Business Unit Level
2. Intermediate I – LGM INTERNATIONAL Government Operations Americas Home Office – Part VIII
3. Intermediate II – LGM INTERNATIONAL Government Operations Global Home Office – Part VIII
4. Intermediate III – LGM INTERNATIONAL Corporate Home Office – Part VIII

These Disclosure Statements which describe LGM INTERNATIONAL Overhead pools are available for review by authorized government agencies. The disclosure statement is proprietary and distribution and use are restricted. The cognizant DCAA and the cognizant CACO have reviewed and approved the disclosure statements.

12.2 Indirect Rates

12.2.1 Calculation of Indirect Rates

LGM INTERNATIONAL's Government Compliance Department is responsible for the preparation, review and evaluation of Indirect cost rates. LGM INTERNATIONAL's rates are prepared, reviewed and evaluated on a quarterly basis for actual/historical costs as well as for forward pricing purposes based on the quarterly reforecast prepared by the LGM INTERNATIONAL Accounting and Finance Department. The reforecast process is a bottom up analysis with input from all responsibility center managers and reviewed by the responsible manager. These reforecasts are then evaluated by a corporate management team. This process is controlled by the Corporate Accounting and Finance Department in accordance with the established plans, templates and review process. Once the reviews are completed, the final budget reforecast is submitted to LGM INTERNATIONAL management for final review and approval, once again in accordance with established accounting and finance procedures.

The Government Compliance Department uses the resulting approved budget and prepares the indirect expense rates based on that data. If the resulting rates exceed the variance allowed in the FPRA, Government Compliance will provide an updated FPRA proposal to the DCMA based on the revised forecast. If the rates variance is within the tolerance provided in the FPRA, the established, negotiated rates will remain in effect. These indirect rate updates are reviewed by both DCMA and DCAA.

Once the rates are approved by the DCMA, the Government Compliance Department notifies the appropriate representatives from Contract Administration, Accounting and Finance and Project Controls functions.

Use or disclosure of data contained on this sheet is
subject to the restriction on the title page of this document.

Page 106

12.2.2 Cost Identification to Pools

LGM INTERNATIONAL's accounting system accumulates and reports the costs associated with the various indirect cost pools by responsibility center (R/C) by the element of cost (labor and other indirect costs). The accumulation of the R/C costs into the indirect expense pools is the responsibility of the Government Compliance Department.

12.2.3 Allocation to Contract

The rates specified in the FPRA are applied to the appropriate cost allocation base. The calculation includes all projects, both commercial and government. The procedures for identifying overhead costs and allocating the pools are reviewed and accepted by the DCAA and the DCMA.

12.2.4 Overhead Application to CWBS

Overhead application is applied at the contract level through the LGM INTERNATIONAL EVMS. Provisional and incurred overhead rates may be applied at any level of the CWBS chosen by the Project Manager. The Control Account coding and CWBS coding structures provide for the summarization of these overhead costs to the total contract level from any point of allocation.

12.3 Base Development

The FPRA is based on LGM INTERNATIONAL's annual Plan and updated for each quarterly Reforecast. The current calendar year budget is constantly monitored and updated quarterly.

12.3.1 Preparation of Forecasts

In the third quarter of each fiscal year, the preparation of the next years Plan begins. The corporate Accounting and Finance Department issues a request to all R/C managers and executives which includes instructions in preparing the individual R/C and revenue budgets and the schedule to completion.

Revenue is forecasted based upon those contracts in which Senior Management and Marketing Management expect to win after the budgeting process has been completed.

For those contracts currently signed, estimates of revenue and cost are prepared for each month based upon contract terms and conditions, work schedules and other known factors.

Based upon past history, estimates are made for those contracts we expect to win. Included in these estimates are contract revenues and costs for "anticipated" contracts, i.e., these already in the proposal process plus those which we know and anticipate will be proposed during the months before the next year begins.

12.3.2 Operating Expenses

The operating forecast for the next year includes costs forecasted at a detailed level. Costs are forecast at the level of activity required to support the revenue projections. Operating expenses are highly labor intensive and the primary costs are salaries and wages, burden and benefits, and other G&A costs related to salaries - such as travel. The other primary costs are facilities and related costs. The budgeting process includes indirect R/Cs as well as project related forecasting.

12.3.3 Corporate Accounting Review of Forecasts

When the forecasts are completed by the managers, Lines and overhead departments, they are reviewed by Corporate Accounting as to being rational and consistent. If they are rejected, they are returned for revisions.

12.3.4 President's Review of Final Forecasts

The final forecasts are forwarded to the President's office, which includes the Executive Vice Presidents of Finance and Operations for LGM INTERNATIONAL.

Use or disclosure of data contained on this sheet is subject to the restriction on the title page of this document.

Page 107

12.3.5 Approved Budget

The final budget is submitted into the accounting system by line item for review monthly. The Government Compliance Department is forwarded the budget to calculate the ensuing year Provisional and Billing Rates for negotiation with the CACO. If the FPRA is adjusted, the resulting rates are incorporated into a "rate matrix" which is provided to the Contracts department, Project Controls staff and Accounting and Finance staff as needed.

12.4 Overhead Budget Control

Each organization which has authority to incur overhead costs receives a copy of their approved R/C budget by element of cost.

Each month, all financial information is processed in the General Ledger Accounting System (General Books of Account), according to charges originated by responsibility centers performing specific, authorized services for the project, i.e., payroll, accounting, etc. Ledger reports detail current month transactions and provides year-to-date summarized totals. All ledger data are adequately supported by auditable documentation which is referenced to provide a traceable audit trail. The reports also reflect variances between actual R/C costs and budgeted or planned costs.

The overhead departments review their variances. Material variances (usually over 15-20%) may require explanations. If required, the R/C manager will forward an explanation to the appropriate accounting and finance manager. The variances may be due to changes in corporate requirements caused by economic conditions, increased business volume or other possible reasons. The responsible manager will take whatever action, within their authority, to correct an adverse situation.

Use or disclosure of data contained on this sheet is subject to the restriction on the title page of this document.

Page 108

13.Data Analysis and Reporting

Use or disclosure of data contained on this sheet is
subject to the restriction on the title page of this document.

Page 109

13.1   Overview

LGM INTERNATIONAL EVMS provides the ability to collect, report, update and analyze cost/schedule information and determine project performance against established baselines throughout the project life-cycle. LGM INTERNATIONAL EVMS provides automated updates of progress and BCWP for customer and internal reports. Variance identification and analysis aid in isolating causes of deviations from the plan. The projection of net effect of such deviations assists in evaluating the project's final performance against contract targets. Through timely reporting, both internal and external, management can take actions necessary to bring variances under control.

This section addresses the process for accomplishing the monthly task of reporting, updating, analyzing and utilizing the performance measurement aspects of LGM INTERNATIONAL EVMS for effective project control.

LGM INTERNATIONAL EVMS provides the capability to:

☐ Produce both internal and external performance measurement reports at any selected level of the Contract Work Breakdown Structure (CWBS) and/or Organization Breakdown Structure (OBS).

☐ Identify and report variances at the Control Account level in terms of labor, material and/or ODC, as appropriate, resulting from the comparison of performance measurement data: BCWS, BAC, EAC, BCWP, and ACWP.

☐ Summarize the data elements and associated variances through the OBS and CWBS to the reporting level specified in the contract. Internal performance reports reconcile to data reported to the customer.

☐ Identify managerial actions taken as a result of this analysis and monitor corrective actions to the point of resolution.

☐ Develop revised estimates of cost at completion for each Control Account with significant variances or with known changes based on performance to date and estimates of future conditions.

☐ Summarize the Control Account estimates of cost at completion for the CWBS elements identified in the contract and compare these with the CBB and the latest statement of funds requirement reported to the customer.

13.2   Data Update

Each month, a routine update of each Control Account is performed to provide the most recent data possible for the performance measurement analysis process. During this updating cycle, BCWP, ACWP and, as appropriate, EAC are provided to the Project Office to reflect performance-to-date, changes to the baseline and projections of future performance.

The CAMs are responsible for statusing BCWP and EAC. The Project Office is responsible for compiling the updated information, checking the input and overseeing the generation of internal system output reports.

BCWP is earned in the same manner BCWS was established. For example, if work was planned (BCWS) using the Apportioned Effort method, BCWP must be earned using the Apportioned Effort technique.

13.3   Monthly Reporting Cycle

The reports generated from LGM INTERNATIONAL EVMS are updated and generated monthly. The large amount of data, the number of people required to make inputs, the processing time and other considerations require that an orderly process be used to collect, review, report and use the data generated by LGM INTERNATIONAL EVMS. Much of the basic data for analysis is financial in nature and thus the reporting cycle is based on the fiscal month accounting calendar (Figure 13-1).

Briefly, the CAM is responsible for providing Control Account status to the Project Office immediately following the close of the fiscal month accounting calendar. This, plus the accumulation of actual costs, allows the monthly performance report to be generated. CAMs receive the portions of the monthly performance report that include their Control Account(s). This allows them to see their performance each month. Figure 13-2 is an example of a Monthly Performance Report.

13.3.1 Control Account Manager Responsibilities

On a monthly basis, the CAM will provide the Project Office with an assessment of progress achieved during the month (i.e., provide a Budgeted Cost of Work Performed (BCWP) input for each Work Package). The cutoff date for this assessment is the last Saturday of the month and is due to the Project Office no later than the following Thursday for labor and material. This update is accomplished by updating and verifying the progress measurement database and posting percent complete to the PMB database. BCWP will be calculated in the same manner as it was planned. The CAM also has responsibility for EAC assessment.

13.3.2 Project Office Responsibilities

The CAM's assessment of monthly accomplishment is then reviewed and input into LGM INTERNATIONAL EVMS. LGM INTERNATIONAL EVMS then provides a CPR by OBS and CWBS element. These reports display current period and cumulative to date BCWS, BCWP, ACWP and their associated Schedule Variance (SV), Cost Variance (CV), plus BAC and EAC and their difference, Variance at Completion (VAC) at the Control Account level. LGM INTERNATIONAL EVMS summarizes all Control Account data in such a way that no Control Account is summarized to more than one CWBS or OBS element. Each CAM receives copies of the CPR and other performance reports by functional organization that displays data for each Control Account. Each Functional Manager also receives copies of the CPR and other performance reports for the appropriate functional area.

13.4 Variance Analysis

13.4.1 Variance Analysis Thresholds

Variance Analysis is performed at the Control Account level and reported at the CPR CWBS and OBS reporting levels. Variance Analysis thresholds are established for current period, inception to date and at completion. Variances are automatically calculated by LGM INTERNATIONAL EVMS. Variances exceeding the threshold levels are flagged and reported to the CAM.

Thresholds are assigned at project level, based on review of the program scope and reporting requirements.

13.4.2 Variance Types

Variance Analysis provides the means to determine the cause for an out-of-balance condition in performance. There are three major types of variances, each reflecting a different view of performance. They are: Cost Variance (CV), Schedule Variance (SV) and At-Completion Variance (VAC).

13.4.2.1 Cost Variance

Cost Variance is the difference between BCWP and ACWP:

$$CV = BCWP - ACWP$$

Since both are tied to the amount of work performed, CV is an accurate assessment of cost performance for the work completed. When CV is positive, (BCWP > ACWP), the work is costing less than budgeted and is favorable. When CV is negative (BCWP < ACWP), the variance is unfavorable.

Through analysis of the above components of a Cost Variance, the CAM and management can better develop action plans and monitor their effectiveness in reducing unfavorable variances.

A cost index, called a Cost Performance Index (CPI), is calculated by dividing BCWP by ACWP:

$$CPI = BCWP/ACWP$$

The index indicates the overall cost efficiency at which the work has been performed. A CPI of greater than 1.0 indicates, for example, that for every dollar of budgeted work performed the actual cost of that work is less. This is a favorable variance.

Use or disclosure of data contained on this sheet is subject to the restriction on the title page of this document.

Page 111

13.4.2.2 Schedule Variance

Schedule Variance is calculated as the difference between BCWP and BCWS:

$$SV = BCWP - BCWS$$

SV indicates schedule performance against the plan in terms of dollars. While CV provides an accurate assessment of cost performance, SV should only be used as an indicator because it does not reflect criticality and does not filter out early starts prior to the planned dates. SV can also be calculated as a percentage of BCWS by dividing SV by BCWS.

A schedule index, called the Schedule Performance Index (SPI), is calculated by dividing BCWP by BCWS.

$$SPI = BCWP/BCWS$$

An SPI of greater than 1.0 indicates a favorable variance.

13.4.2.3 Variance at Completion

VAC is calculated by subtracting the EAC from the BAC:

$$VAC = BAC - EAC$$

When VAC is positive (BAC > EAC), it indicates a forecast under run. This is a favorable variance. Negative values indicate a forecast overrun. VAC can be calculated as a percentage of the BAC to assess the variance for significance. In the early stages of a project, significant VAC is uncommon because VAC is small. This is because there is not enough work completed to assess the total impact on the project.

TCPI is cost efficiency that must be achieved in the remaining period of performance to complete the total work scope within the budget (BAC) or cost (EAC) objective. A To-Complete Performance Index (TCPI) is calculated by dividing the budget for the remaining work (BAC - cumulative BCWP) to be accomplished by the estimated cost to accomplish the remaining work (EAC - cumulative ACWP).

$$TCPI = (BAC - BCWP)/(EAC - ACWP)$$

13.4.3 Variance Reporting

Upon receiving their Monthly Performance Reports, CAMs note those Control Accounts that have exceeded the variance threshold. These Control Accounts will need a more thorough review to determine the cause of the variance and require a VAR (Figure 13-3).

VARs may be due to one variance, an SV, CV or VAC, or all three. Whether the variance is favorable or unfavorable, the CAM is required to review and determine the cause of the variance, assess its impact and develop a corrective action plan. When this plan is submitted to the Project Office, it is recorded on a Corrective Action Log (Figure 13-4). The Corrective Action Log is used by management to remain aware of problem areas, the corrective actions taken, and the results of the corrective action.

CAMs are responsible for those elements of cost under their control. Overhead costs are monitored and analyzed by the Finance and Accounting office. The CAM is responsible for labor, material, subcontract and ODCs.

VARs are developed at the CAM (Control Account), Functional Manager (Organization) and Project levels (total project). This is to ensure all Project participants are kept aware of superior performance as well as problem areas.

VARs are also used in developing an update to the EAC. The CAMs must take these impacts into account when developing their ETC.

13.5 Estimate at Completion

The primary objective in preparing an EAC is to provide an accurate projection of cost at contract completion.

Use or disclosure of data contained on this sheet is subject to the restriction on the title page of this document.

Page 112

The EAC is calculated as the cumulative ACWP plus the ETC. The cumulative ACWP portion of the EAC is the more objective portion of the EAC formula: nevertheless, it should be reviewed with respect to missing or incorrect charges. As a minimum, EACs are based on performance to date, actual costs to date, the projections of future performance, economic escalation, expected Overhead/G&A rates and material commitments.

The basis of the ETC is the estimate for the remaining authorized scope of work. Using this as a starting point, the CAM must evaluate its current validity as estimated future costs. The recovery plan for any existing cost and schedule variances will have an impact on the ETC and must be considered. The CAM must also consider any past problem areas that may have an effect on future work, technical breakthroughs, improvements in methods and any other knowledge regarding the tasks to be done.

EACs are developed with varying degrees of detail and supporting documentation. On a monthly basis, the CAM evaluates the existing EAC for accuracy. On a monthly basis, the CAM will review the forecast value to complete the project, ETC, and confirm that this is the EAC that will be required to complete the project. Figure 13-5, is an example of a Control Account ETC planning sheet. This EAC is less detailed than the comprehensive EAC and is often called a "current performance EAC" or "EAC update."

13.5.1 Comprehensive EAC

A comprehensive EAC is prepared at least annually and whenever current performance indicates the current project EAC is invalid. A comprehensive EAC may be done in concert with the ETC process. The Project Manager issues a memorandum directing all CAMs and Functional Managers to prepare a comprehensive ETC. This memo will include the effective date, copies of the current Intermediate Schedules, due dates and any assumptions/facts to be used in developing the ETC.

The Project Office issues the comprehensive EAC memorandum to all Functional and CAMs. While the CAMs are preparing their ETC, the Project Office reviews the contract scope, associated CTC, and the proposed cost for AUW. They also verify the status of outstanding commitments for material, major subcontracts and ODC to ensure they are complete and do not duplicate actuals already recorded in the Cost Ledger. The Project Office requests Finance to provide the expected future Overhead/G&A rates by pool and fiscal year. These rates will be used to develop the Overhead/G&A cost ETC.

ETCs developed by the CAMs are forwarded to their Functional Managers for approval. Approved ETCs are forwarded to the Project Office. Disapproved ETCs are returned to the CAM for iteration.

Upon receipt of the ETCs from the CAMs/Functional Managers, the Project Office reviews them to ensure that they are reasonable and consistent with the remaining contract scope of work. Disapproved ETCs are returned to the Functional Manager for iteration by the CAM. Approved ETCs are summarized and the appropriate Overhead/G&A rates are applied by the Project Office. All ETCs are added to the cumulative ACWP to determine the EAC.

13.5.2 EAC Updates

Each month, upon receipt of the monthly performance report, the CAM reviews the performance and EAC. If the EAC remains consistent with the performance to date and no additional information is available which would make the EAC unreasonable, no action is required by the CAM (apart from completing a VAR, if required). If the CAM no longer believes the current EAC, a new time phased ETC by element of cost is developed. The time-phased ETC is provided to the Project Office.

Upon completion of the Control Account ETC/EAC, the CAM forwards the estimate to the Functional Manager for review and approval. If approved, the Functional Manager will sign off the estimates and forward them to the Project Office. If the Functional Manager disagrees, the package will be annotated indicating the areas of disagreement and returned to the CAM. The CAM and Functional Manager will negotiate until a valid ETC/EAC is obtained.

During this same period, the Functional Managers will review and approve the Control Account VAR's. Both the VARs and the revised ETCs/EACs are submitted to the Project Office. If required, the Functional Managers will prepare a functional

Use or disclosure of data contained on this sheet is subject to the restriction on the title page of this document.

Page 113

VAR when their organization/department have exceeded established reporting thresholds. They will be notified by the Project Office when a functional VAR is required.

The VARs prepared by the CAMs and the Functional Managers address EAC direct costs only. Finance performs, on a monthly basis, an analysis of the Overhead/G&A costs and informs the Project Office via a formal analysis when they forecast the Overhead/G&A rates to vary from the planned rate and thus impact the project EAC.

In the case of fast-track programs (where design overlaps significantly with construction/installation), Procurement/Materials may act as CAM for early procurement activities. In these cases, Procurement/Material will provide EAC for these materials based on commitments and provide this input to the Project Office.

When the Project Office receives the ETC/EAC package from the Functional Managers and Procurement/Materials as appropriate, they evaluate it against past performance indicators and their overall knowledge of the project. If the Project Office concurs, they update the project EAC including any Overhead/G&A cost modifications. The Overhead/G&A cost EAC will be based upon the time-phased direct cost EAC by Control Account and the applicable future Overhead/G&A rates furnished by Finance. The total project EAC is forwarded by the Project Office to the Project Manager for approval.

13.6 Performance Reporting

13.6.1 Internal Reporting

LGM INTERNATIONAL EVMS provides the flexibility to report the same performance data in different formats. The same BCWS, BCWP and ACWP can be summarized in different ways because each Control Account can be traced back to both a CWBS element of work and a functional organization. Control Account performance data can be reported by CWBS, summarizing to each level of the CWBS without one element of cost summarizing to two different higher CWBS levels. This same performance data can be summarized by functional organization according to the OBS, from the Control Account up to the highest organizational levels without one element of cost summarizing to two different higher OBS levels. EACs are available for each level of summary.

Typically, the monthly performance data are also plotted so that ACWP and BCWP can be compared and BCWP and BCWS can be compared (Figure 13-6).

13.6.2 Customer Reporting

The primary customer reports are the CPR, C/SSR, and the CFSR. Their frequency and requirement are determined by the customer. LGM INTERNATIONAL management uses the same reports submitted to the customer for internal review and analysis.

13.6.2.1 Cost Performance Report

The CPR is a formally submitted report showing project performance consisting of five separate formats:

- Format 1 — Cost Performance by WBS (Figure 13-7)
- Format 2 — Cost Performance by Organizational Category (Figure 13-8)
- Format 3 — Cost Performance by Baseline (Figure 13-9)
- Format 4 — Cost Performance by Staffing (Figure 13-10)
- Format 5 — Cost Performance by Explanation and Problem Analysis (Figure 13-11)

Report Formats 1 through 4 are standard outputs from the LGM INTERNATIONAL EVMS database. Format 5, the Problem Analysis Report, is a narrative discussion of contract problems as identified in the preceding four formats. The month-end cycle includes a review by the project organization prior to finalization for issue to the customer. Figure 13-1 illustrates this basic monthly reporting cycle.

Use or disclosure of data contained on this sheet is
subject to the restriction on the title page of this document.

Page 114

The CPR is a composite report generated by the Project Office for the customer. The actual number and detail of the CPR formats are established with the customer at the time of contract definition or negotiation.

The first two formats of the CPR are generated by WBS and by organization. The elements of BCWS, BCWP, ACWP, BAC and EAC are verified for correctness and then included in the CPR.

Format 3, Baseline Plan, is prepared from LGM INTERNATIONAL EVMS database, which contains BCWS for the period of performance of the contract.

Format 4, Personnel Loading, is prepared by the Project Office from the LGM INTERNATIONAL EVMS database, which includes ETC data received from the functional organizations that are performing on the contract.

13.7   Performance Measurement, Reporting, and Analysis Flowcharts

The flowchart in Figure 13-14 illustrates the process of measuring performance, generating internal performance reports, and analyzing progress. The flowchart in Figure 13-15 illustrates the process of generating the customer's CPR. Both flowcharts are described in the following paragraphs.

13.7.1   Performance Measurement, Analysis, and EAC Update Flowchart

The flowchart in Figure 13-14 illustrates the process of measuring performance, generating internal performance reports, and analyzing progress. Paragraphs H through Y below describe the monthly process for updating the EAC.

A.   Upon contract award, Business Development/Contracts, as agent for the Product Line Vice President, issues a CWA authorizing the Project Manager to expend LGM INTERNATIONAL resources to accomplish the contract effort.

B.   The Project Manager reviews the variance analysis thresholds set forth in the contract and establishes internal thresholds for reporting.

C.   The internal thresholds are communicated to Functional Management via a Project Directive.

D.   The Project Manager authorizes the Functional Managers who in turn authorize the CAMs via PWAs and CAWAs, respectively. The CAMs begin work as authorized.

E.   Each month, the CAM updates the CAP, annotating schedule accomplishment and projections. The statused CAP is forwarded to the Project Office. Depending on the nature of the work, automated progress measurement systems may be in place for reporting progress at the CAP level. In these cases, the automated progress subsystem will provide all or part of the statused CAP.

F.   Based on the statusing displayed on the CAP and/or from the progress measurement subsystems, BCWP is posted to the LGM INTERNATIONAL EVMS database. Concurrently, with the determination of BCWP, actual costs are entered/accumulated, as appropriate. They are entered into the LGM INTERNATIONAL EVMS at the Control Account or Work Package level.

G.   LGM INTERNATIONAL EVMS is now able to generate Performance Reports. Monthly Performance Reports are generated (at the Work Package level) for all Control Accounts. VARs are generated for all Control Accounts and CWBS and OBS elements that exceed the thresholds established by the Project Manager. ETC reports reflecting the most recent ETC updates are also printed.

H.   Upon receiving these reports, CAMs review their performance.

I.   In reviewing performance, the CAM identifies significant Work Package variances which have contributed to/or offset Control Account level variances. Note that the CAM reviews performance whether or not required to complete a VAR. It is possible that significant variances exist at the Work Package level which offset one another, resulting in acceptable performance at the Control Account level. Identifying these lower level variances and taking appropriate action can solve some problems before they become major concerns.

J.   The CAM also reviews the variances by element of cost. Such a review can yield information not gleaned from a Work Package analysis. Several material Work Packages may be experiencing variances that suggest problems with material usage or with a specific vendor, for example. Conversely, significant labor variances may suggest that improperly trained or supervised personnel have been assigned to the task.

K. The CAM also reviews the schedule position against key or significant milestones. In addition to identifying the cause of existing variances, this review can identify potential schedule slippages that were not recognized when the CAP was statused and reveal that additional resources (and added costs) will be required to maintain schedule. The CAM analyzes any schedule impacts and, using the VAR form, communicates corrective action designed to minimize the impact and regain the baseline schedule. Finally, the CAM reviews any additional supporting data available such as drawing counts, material commitment, etc.

L. When significant variances are experienced, the CAM develops a corrective action plan. The corrective action plan may be formal or informal, depending on the significance of the Control Account level variance (i.e., whether or not a VAR is required). The corrective action should state what is to be done, when it is to be accomplished, and who is responsible for it.

M. The CAM also reviews the planning of unopened work packages. If current variances or other changes which have occurred will impact this future planning, the CAM revises the planning if possible. Section 14 discusses changes to baseline plans in more detail.

N. The CAM also reviews the time phased resource projections in light of all other known information. Considerations include whether existing variances will further impact the EAC, trends and cumulative to date performance, schedule projections and resource availability, and any other information available in assessing the EAC. If appropriate, the ETC projection is revised by marking up the ETC report.

O. If required, the CAM documents the findings of steps H through N on a VAR.

P. The VAR is forwarded to the Functional Manager for review and approval. If the CAM has changed the EAC, the change is attached to the VAR. The Functional Manager reviews the VAR's. Upon agreement with the analysis, corrective action, and EAC, the Functional Manager signs the VARs and forwards them to the Project Manager. If disapproved, they are returned to the CAM for iteration.

Q. Upon receipt of the VARs from the Functional Managers, the Project Office reviews them. The analysis, corrective action, and EAC are also reviewed.

R   If the Project Manager disapproves the VARs, they are returned for iteration. Copies of approved VARs are also returned to the CAM. ETC reports that reflect approved changes to the EAC are forwarded to the Project Office for incorporation to the LGM INTERNATIONAL EVMS database.

S. The Project Office Scheduling Section reviews the VARs and previously received CAPs for schedule impact. Intermediate and the PMS are updated appropriately.

T. Upon receipt of the approved VAR, the CAM implements the identified corrective action.

U. The Project Office maintains a Corrective Action Log of all significant corrective actions annotated and approved on VAR's. The Project Manager monitors these actions to resolution.

V. The Project Office summarizes the approved direct estimated costs for the EAC. The appropriate Overhead/G&A rates are applied to the approved, summarized direct estimated costs to identify the total EAC for the Performance Measurement Baseline.

The Project Manager reviews the MR. The amount of MR forecasted to be consumed before the end of the contract is added to the EAC for the Performance Measurement Baseline to develop the total project EAC.

13.7.2   Customer Reporting Flowchart

The flowchart in Figure 13-15 illustrates the process of preparing the CPR. The following paragraphs describe the flowchart. This flowchart begins at the generation of the internal performance reports (Step G in Figure 13-14). Prior to this time, the CAM is formally authorized to work on the contract, begins work, charges actuals for the work performed, and statuses the CAP to indicate work accomplished. This information is reviewed and input into the LGM INTERNATIONAL EVMS database by the Project Office, which also generates the internal performance reports.

A. Internal performance reports are generated by Project Controls and provided to the Project Manager, Functional Managers, and CAMs. Financial reports are received by Corporate Compliance and Project Finance organizations.

B. Based on the variance thresholds established by the Project Manager (see Steps B and C in Figure 13-14), significant variances are identified and VARs are generated.

Use or disclosure of data contained on this sheet is subject to the restriction on the title page of this document.

Page 116

C. CAMs analyze their performance by reviewing Work Package level data. (See Steps through O in Figure 13-14 for a detailed discussion of this analysis.)

Functional Managers also review their performance and analyze any significant variances. This is normally accomplished by looking at data at the Control Account level and then reviewing identified problems with the appropriate CAMs (see step D, below).

D. CAMs document the cause of significant variances, their impact, and any available corrective action on VARs. These reports are forwarded to their Functional Manager for review and approval. The Functional Manager uses the information provided in the VARs to generate and support the functional level VAR. Control Account VARs, approved by Functional Managers (see Step P, Figure 13-14) and functional level VARs are forwarded to the Project Office.

E. Concurrent with the preparation of VARs dealing with direct cost variances, Corporate Compliance reviews the status of prepared Overhead/G&A cost reports that explain the differences between budgets and actuals. Project Finance prepares the CFSR, normally quarterly, for customer reporting using data from the Accounting and Material Planning and Control subsystems and EAC information from Project Controls to project future funding requirements. These reports are also provided to the Project Manager.

F. The Project Office performs an initial review of the VARs to ensure that they adequately explain the problem, and fully identify what corrective action is planned and who is responsible for it.

G. Those VARs which are inadequate are returned to the functional organizations for iteration. Acceptable VARs are forwarded to the Project Manager for approval. (Note that the Project Manager may also return a VAR to the appropriate Functional Manager if it is felt that more information is necessary).

H. Throughout the Variance Analysis Reporting cycle, the Project Office summarizes data for the Project Manager to use during the monthly Project Reviews. They also perform cursory audits of the data to ensure that it is valid. Where data is found to be significantly in error, corrections are made to data provided to the customer. Minor errors are noted for correction during the following reporting period.

I. The Project Office prepares the external CPR.

Data for Formats 1 and 2 are generated from the LGM INTERNATIONAL EVMS database by summarizing Control Account level data through the CWBS and functional organization, respectively.

The baseline format, Format 3, is generated from the LGM INTERNATIONAL EVMS database. Differences from the previous reporting period can be traced through the CBB, UB and MR logs. Format 4, Staffing, is prepared based on the CAM's ETC. Finally, Format 5, the narrative analysis, is generated by summarizing the VARs provided by the Control Account and Functional Managers.

J. The CPR and CFSR (when required) are reviewed and approved by the Project Manager prior to submitting it to the customer.

The Project Manager conducts a Project Review, normally during the second week after the end of the accounting period - generally during the same time that Variance Analysis Reporting is being accomplished. This allows the Project Manager to interact with Control Account and Functional Managers while the reports are being written. This avoids multiple iterations of the VARs by providing an opportunity to discuss corrective action before developing the final plan. It also provides an opportunity to ensure that corrective action identified in previous months is being properly implemented.

The Project Manager and his or her staff (i.e., the Project Office) maintain a Corrective Action Log of all significant actions identified in the VARs and/or during the Project Review.

13.7.3 Comprehensive EAC Update Flowchart

Figure 13-16 illustrates the process for developing the comprehensive EAC.

A. The Project Manager identifies the need for accomplishing a comprehensive EAC. A comprehensive EAC will be done at least annually or more often if contract performance or other factors indicate a need for it.

Use or disclosure of data contained on this sheet is
subject to the restriction on the title page of this document.

Page 117

B. The Project Manager and Project Controls work together to identify the cost, schedule, technical and administrative ground rules to be used as the basis for developing the comprehensive EAC. For example, the ground rules should include, as a minimum, the following information:

1) Contract Baseline (i.e., the EAC should consider all contractually authorized work through a specific contract change). If the contract is in a state of significant change, specific contract changes may need to be identified as being included and/or excluded from the EAC process.

2) Schedule Forecast. A schedule should be identified (by revision number and date) for use in developing the EAC.

3) Technical issues such as design decisions, additional test parameters, etc., should be spelled out to ensure a common understanding from the start of the EAC process (thus avoiding additional effort in revising the EAC developed by the CAM).

4) Cost/Funding issues. While the EAC should not in any way be "capped" or limited, cost and/or funding issues should be identified and the schedule impacts reflected in the schedule forecast used for the ETC. The month-end on which the EAC should be based should also be defined.

5) Administrative issues, such as a prescribed format for submitting the EAC, additional approval authorities or cycles, etc., should be identified in the ground rules.

C1 & C2 The Project Manager and Project Controls work together to identify the schedule for development, submittal, approval and implementation of the EACs. A comprehensive EAC normally takes 2-3 months to develop, fully approve and implement.

D. Project Controls creates a Directive which documents the results of steps B and C.

E. The Directive is signed by the Project Manager and issued initially to Project Controls and ultimately to the Functional Managers and CAMs.

F. Project Controls receives the Directive.

G. Project Controls adds additional guidance, as appropriate (schedule for computer support, for example) to the Directive. They then forward the Directive and additional guidance to the Functional Managers.

H. The Functional Managers receive the Directive and guidance from Project Controls.

I. The Functional Manager adds any functional specific guidance, such as functional schedule references and internal schedule for review and approval of the EACs. Finally, the Directive and additional guidance (both from Project Controls and the Functional Manager) are forwarded to the CAM.

J. The CAMs receive the Project Directive and additional guidance and begin to develop their comprehensive EACs.

K. The CAM evaluates the performance measurement data for accuracy. This includes the following two steps (L and M below).

L. The CAM first looks at the BCWP to ensure that (1) earned value for all work accomplished has been properly taken, and (2) the earned value technique (BCWP method) has not caused a distortion in the BCWP. An accurate assessment of BCWP significant distortion is made. Note that such an assessment should have the following characteristics:

1) Be documented with rationale.

2) Be as objective as possible.

3) Be consistent with the way the work was planned and BCWS was established. For example, if the CAM planned a task to be worth $500, the value of that task cannot be changed. If, however, the percent complete BCWP method (which allows claim of only up to 80% complete until all work associated with the task has been finished) and the work is actually 95% complete, the assessment should be documented and the ETC based on the remaining 5%. (Note that the BCWP in LGM INTERNATIONAL EVMS will not be increased above the allowable 80% until the Work Package is complete.)

M. The CAM then reviews ACWP for accuracy. This includes ensuring that all charges are appropriate and correctly recorded, and that any previously identified corrections have been properly made. If errors exist, the correct ACWP value is identified, and the CAM completes the appropriate forms to have any errors corrected through the accounting process.

N. The CAM then reviews existing cost and schedule variances, identifying the causes of these variances and reviewing corrective action that has been implemented. Have variances resulted from a single technical issue that has been resolved, or are they a result of on-going and/or growing problems that have not yet been solved? The analysis of existing variances includes the process described in steps O through Q below.

O. The CAM reviews schedule variances to identify their cause(s) and the relationship between them and existing cost variances. (Their impact on future effort and schedule and cost performance will also be considered - see step W.)

P. For labor cost variances, the CAM must review existing wage rate variances as well as labor productivity variances. Material price and usage variances are reviewed for material cost variances.

Q. After reviewing current and cumulative variances, the CAM reviews performance trends. The EAC backup file will be used as additional input for this process (Section 14). Graphs of performance measurement indicators may offer insight into trends that may otherwise be overlooked.

R   Finally, the CAM begins the process of evaluating the remaining effort and estimating the costs associated with completing this effort. This includes both an evaluation of the effort itself and a projection of cost and schedule performance that can be expected in accomplishing that effort. This process is further described in steps S through V.

S. The technical risk associated with the remaining effort should be reviewed in relationship with the risk associated with completed effort. This assessment is critical to successfully projecting performance for remaining effort when using experienced performance as the starting point or basis. In other words, is the remaining effort less technically challenging than the effort already completed? If so, it may be reasonable to expect improved cost and schedule performance.

T. The CAM reviews the current schedule projections in light of the effort remaining in the Control Account and the schedule accomplishment date. If appropriate, the projections are redlined to more accurately reflect the way the remaining effort will be accomplished.

U. To accurately develop an ETC for material, the CAM reviews all material use and price differences.

Are the original estimates for this material still accurate or have prices changed? Are the quantities identified still accurate or have design changes or usage variances caused a need for more or less material? Input on material commitments may also be provided by the Project Procurement/Material Group.

V. Finally, based on current schedules and information, the CAM evaluates the planning for tasks (Work/Planning Packages and/or Control Accounts, as appropriate) which have not yet started. Where appropriate, this planning is revised (see Section 14) or new schedule projections are made.

W. Based on the knowledge gained during the process of (1) evaluating performance measurement data for accuracy, (2) evaluating existing variances, and (3) evaluating future effort and anticipated performance (steps K through V), the CAM develops a time-phased ETC for each Control Account.

X. Working with the Project Office, the CAM identifies and applies the appropriate labor rates to the time-phased labor hours. The dollarized ETC is now added to the ACWP to develop the EAC for labor.

Y. The Functional Manager reviews the CAM's ETC/EAC and all supporting documentation for validity and reasonableness.

Z. If the Functional Manager agrees with the CAM's estimates, all documentation is forwarded to Project Controls. If there is a disagreement, the documentation is returned to the CAM for revision where necessary.

AA. The Functional Manager also supports Project Controls in developing an ETC for all effort which has not yet been budgeted to Control Accounts (i.e., those tasks associated with UB).

BB. While Control Account and Functional Managers are developing and reviewing Control Account and functional estimates at completion, Project Controls develops estimates at completion for project level elements. This process is described in steps CC through EE, below.

CC. Project Controls evaluates and develops an ETC for the Project Manager's MR. This estimate identifies how much of the reserve is anticipated to be issued as budget for newly identified tasks prior to contract completion.

Use or disclosure of data contained on this sheet is
subject to the restriction on the title page of this document.

Page 119

DD. Estimates are also developed for Overhead/G&A costs. Section 12, describes in detail the process for budgeting, applying ACWP, and developing estimates for Overhead/G&A costs. During the comprehensive EAC process, the current estimates are reviewed for accuracy. If necessary, support in revising these estimates is requested from Project Finance.

EE. Project Controls has the responsibility of validating the EACs developed by the various organizations. This involves both reviewing the supporting documentation to ensure that all necessary information has been provided and performing statistical analysis. This includes a comparison of the CPI to the TCPI.

Unless the Project Manager has directed otherwise in the EAC ground rules, the purpose of the comprehensive EAC validation process is to provide the Project Manager with additional information and analysis, not to evaluate an ETC/EAC that has been generated by a Control Account or Functional Manager as acceptable or unacceptable. This responsibility rests with the Project Manager and technical staff. Normally, Project Controls has authority to reject comprehensive EACs only on the basis of their being incomplete (i.e., not considering all remaining effort) or not completed in accordance with the established ground rules.

FF. Project Controls consolidates all comprehensive EACs into summarized form(s) and forwards them to the Project Manager for review and approval.

GG. The Project Manager reviews the comprehensive EACs and results of Project Controls validation analysis.

HH. The Project Manager approves or disapproves the comprehensive EACs. Those not approved are returned to the appropriate organization (i.e., the organization which generated the estimate and is responsible for it) for rework. Approved EACs are forwarded to Project Controls.

II. Project Controls enters approved EACs into LGM INTERNATIONAL EVMS for incorporation in the next report submitted to the customer.

JJ. Project Controls also returns the approved EAC packages to the responsible CAM.

KK. The CAM receives the approved EAC package for incorporation with the Control Account documentation.

Use or disclosure of data contained on this sheet is subject to the restriction on the title page of this document.

Page 120

PROJECT OFFICE ACTIONS:

Complete and Submit Customer Reports

Run Final Direct Material, Sub, ODC Reports

Run Preliminary Budget and Performance Measurement Reports

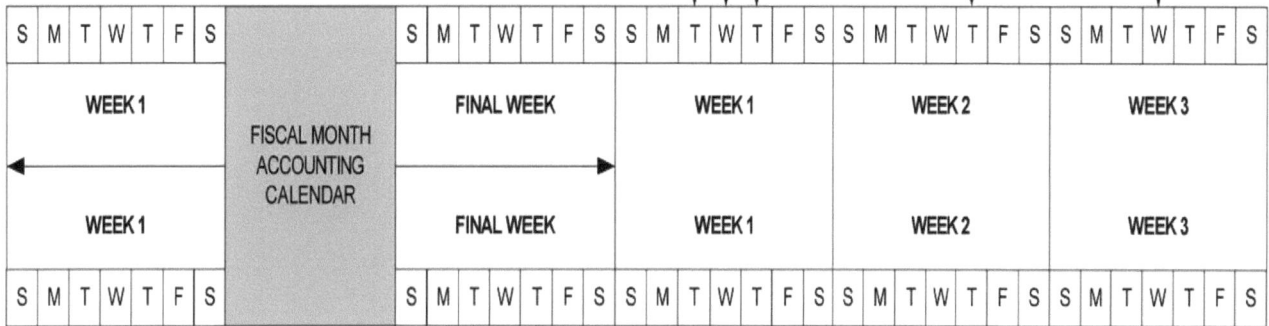

Run Month End Direct Labor Reports

Run Preliminary Budget and Performance Measurement Reports

S	M	T	W	T	F	S		S	M	T	W	T	F	S	S	M	T	W	T	F	S	S	M	T	W	T	F	S	S	M	T	W	T	F	S

WEEK 1	FISCAL MONTH ACCOUNTING CALENDAR	FINAL WEEK	WEEK 1	WEEK 2	WEEK 3
WEEK 1		FINAL WEEK	WEEK 1	WEEK 2	WEEK 3

S	M	T	W	T	F	S		S	M	T	W	T	F	S	S	M	T	W	T	F	S	S	M	T	W	T	F	S	S	M	T	W	T	F	S

CONTROL ACCOUNT MANAGER ACTIONS:

Submit All Approved Work Package Planning Revisions

Submit All Control Account and Schedule Status Updates

Submit Labor and Materials, ETC., EAC Updates

Complete and Submit Control Account Variance Analysis Rpeorts

Figure 13-1   Monthly Reporting Cycle

Use or disclosure of data contained on this sheet is subject to the restriction on the title page of this document.

Page 121

Project: ANCDF
Date: 21 Aug 03

Report: CPR_W
Page: 1

Project XYZ

CPR Working Format

Code Description		CURRENT PERIOD					CUMULATIVE TO DATE					AT COMPLETE		
		BCWS	BCWP	ACWP	SCH VAR	COST VAR	BCWS	BCWP	ACWP	SCH VAR	COST VAR	BAC	EAC	VAR
01	Elevated Slabs	0	0	0	0	0	0	0	0	0	0	105	105	0
02	Civil Structures	16	50	88	34	-38	16	50	88	34	-38	1050	1050	0
14 31 3	CHB Concrete	16	50	88	34	-38	16	50	88	34	-38	1155	1155	0
01	Main Structure	0	0	0	0	0	0	0	0	0	0	550	550	0
02	Misc.	0	0	0	0	0	0	0	0	0	0	150	150	0
14 31 4	CHB Structural Steel	0	0	0	0	0	0	0	0	0	0	700	700	0

Figure 13-2   Sample Monthly Performance Report

Use or disclosure of data contained on this sheet is
subject to the restriction on the title page of this document.

Page 122

COST ACCOUNT
- ☐ OBS   TITLE/DESC._____NO. _____
- ☐ CAM
- ☐ FUNCTIONAL MOR   NAME:_____REPORTING PERIOD:
                                    MO:_____YR. _____

	BCWS	BCWP	ACWP	SCH-VAR		COST-VAR	
				$	%	$	%
REPORTING PERIOD							
CUMULATIVE-TO-DATE							

BASIS OF VARIANCE	BAC	EAC	VAR-AT-COMP.	
			$	%
☐ SCH VARIANCE ☐ COST VARIANCE ☐ VARIANCE AT COMPLETION				

VARIANCE CAUSE:

COST ACCOUNT/PROGRAM IMPACT:

CORRECTIVE ACTION

CORRECTIVE ACTION ASSIGNED TO:

CORRECTIVE ACTION COMPLETION DATE:

COST ACCOUNT MANAGER:_____DATE: _____ | CORRECTIVE ACTION LOG NO.

FUNCTIONAL MANAGER_____DATE:

PROGRAM OFFICE: _____ DATE:

Figure 13-3   Sample Variance Analysis Report

PROJECT

PAGE:
CUSTOMER:

LOG NUMBER	DATE ASSIGNED	ASSIGNED TO	ACTION	DISPOSITION	CLOSURE DATE

Figure 13-4   Sample Corrective Action Log

Report: CAP LJ	Control Account ETC Planning Sheet	Page : 4

Project: ANCDF	Description: PROJECT XYZ	Approval: Project Office _____
Run Date: 08/03/03	Status Date: 07/31/03	Function Manager _____ Control Account Manager _____

Control Account: 14313	Description: CHB CONCRETE	CA Manager: SMITH
Scheduled Start: 07/15/03	Scheduled Finish: 05/25/04	Status: Open

Work Package: 02	Description: CIVIL STRUCTURES		Scheduled Start: 07/15/03		Scheduled Finish: 05/25/04							

PMT: Milestone	Status: Open	To-date	Aug 03	Sep 03	Oct 03	Nov 03	Dec 03	Ja n04	Feb04	Mar 04	Apr 04	May 04	At Complete
Cost Element FND/CIVIL MATERIALS	ETC	58	28	39	55	71	110	83	50	50	28	20	592
	BCWP	50	0	0	0	0	0	0	0	0	0	0	
	ACWP	58	0	0	0	0	0	0	0	0	0	0	
LABOR	ETC	41	23	36	49	66	102	76	45	45	25	19	527
	BCWP	45	0	0	0	0	0	0	0	0	0	0	
	ACWP	41	0	0	0	0	0	0	0	0	0	0	
Work Package Totals	ETC	99	51	75	104	137	212	159	95	95	53	39	1119
	BCWP	95	0	0	0	0	0	0	0	0	0	0	
	ACWP	99	0	0	0	0	0	0	0	0	0	0	

Figure 13-5   Sample Control Account ETC PLANNING Sheet

Dollars, in Millions

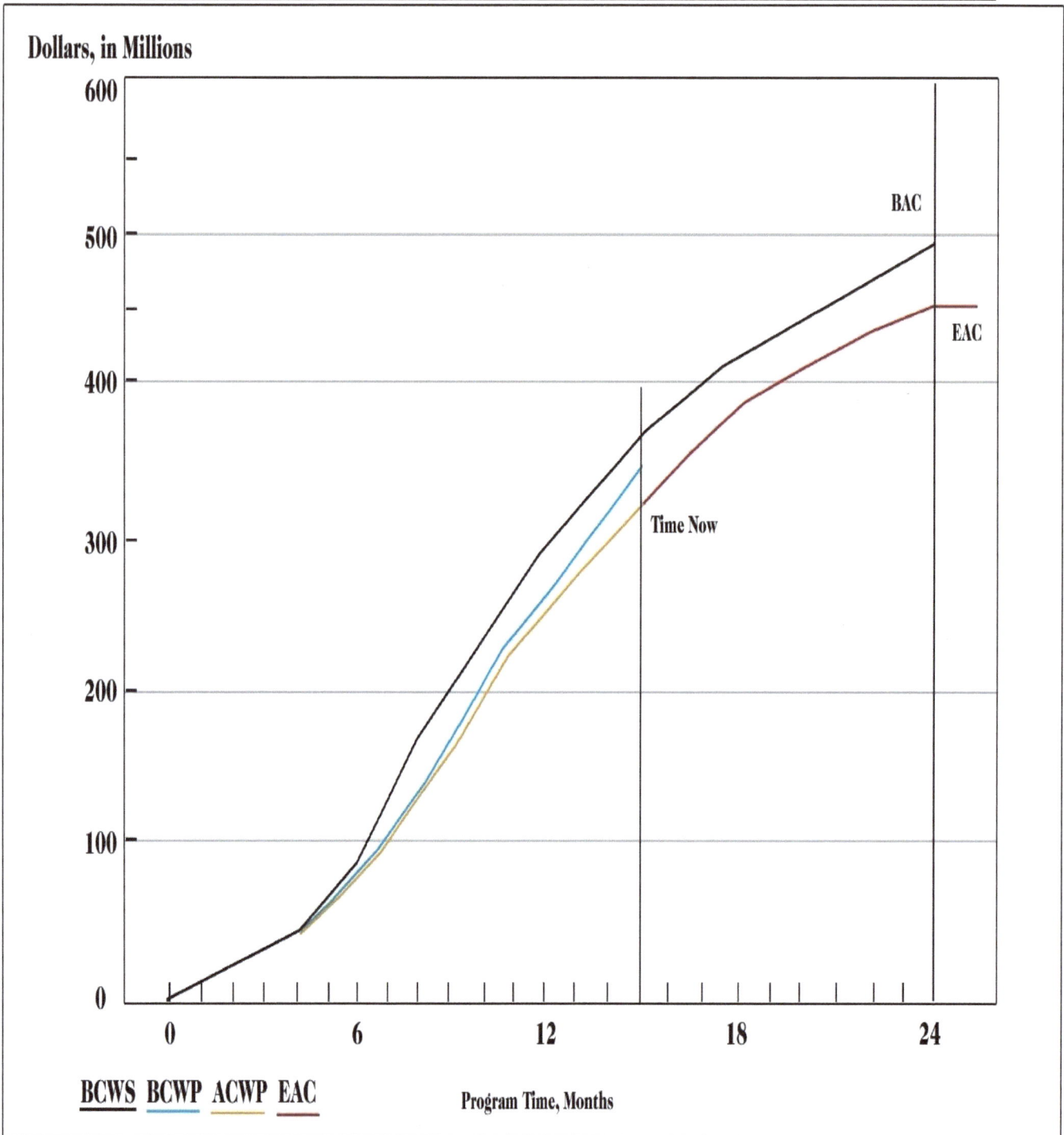

Figure 13-6   Monthly Performance Data

Use or disclosure of data contained on this sheet is
subject to the restriction on the title page of this document.

Page 126

COST PERFORMANCE REPORT FORMAT 1 - WORK BREAKDOWN STRUCTURE	DOLLARS IN _____	Form Approved OMB No. 0704-0188

1. CONTRACTOR
a. NAME
b. LOCATION *(Address and ZIP Code)*

2. CONTRACT
a. NAME
b. NUMBER
c. TYPE
d. SHARE RATIO

3. PROGRAM
a. NAME
b. PHASE *(X one)* RDT&E PRODUCTION

4. REPORT PERIOD
a. FROM *(YYYYMMDD)*
b. TO *(YYYYMMDD)*

5. CONTRACT DATA

a. QUANTITY	b. NEGOTIATED COST	c. EST. COST AUTHOR-IZED UNPRICED WORK	d. TARGET PROFIT/ FEE	e. TARGET PRICE	f. ESTIMATED PRICE	g. CONTRACT CEILING	h. ESTIMATED CONTRACT CEILING

6. ESTIMATED COST AT COMPLETION

	MANAGEMENT ESTIMATE AT COMPLETION (1)	CONTRACT BUDGET BASE (2)	VARIANCE (3)
a. BEST CASE			
b. WORST CASE			
c. MOST LIKELY			

7. AUTHORIZED CONTRACTOR REPRESENTATIVE
a. NAME *(Last, First, Middle Initial)*
b. TITLE
c. SIGNATURE
d. DATE SIGNED *(YYYYMMDD)*

8. PERFORMANCE DATA

ITEM (1)	CURRENT PERIOD					CUMULATIVE TO DATE					REPROGRAMMING ADJUSTMENTS		AT COMPLETION		
	BUDGETED COST		ACTUAL COST WORK PERFORMED (4)	VARIANCE		BUDGETED COST		ACTUAL COST WORK PERFORMED (9)	VARIANCE		COST VARIANCE (12)	BUDGET (13)	BUDGETED (14)	ESTIMATED (15)	VARIANCE (16)
	WORK SCHEDULED (2)	WORK PERFORMED (3)		SCHEDULE (5)	COST (6)	WORK SCHEDULED (7)	WORK PERFORMED (8)		SCHEDULE (10)	COST (11)					
a. WORK BREAKDOWN STRUCTURE ELEMENT															
b. COST OF MONEY															
c. GENERAL & ADMINISTRATIVE															
d. UNDISTRIBUTED BUDGET															
e. SUBTOTAL *(Performance Measurement Baseline)*															
f. MANAGEMENT RESERVE															
g. TOTAL															

9. RECONCILIATION TO CONTRACT BUDGET BASE

a. VARIANCE ADJUSTMENT															
b. TOTAL CONTRACT VARIANCE															

DD FORM 2734/1, AUG 96 PREVIOUS EDITION MAY BE USED. LOCAL REPRODUCTION AUTHORIZED.

Reset

Figure 13-7   Cost Performance Report, Format 1—Work Breakdown Structure

Use or disclosure of data contained on this sheet is
subject to the restriction on the title page of this document.

Page 127

| COST PERFORMANCE REPORT
FORMAT 2 - ORGANIZATIONAL CATEGORIES   DOLLARS IN _____ | *Form Approved*
OMB No. 0704-0188 |

The public reporting burden for this collection of information is estimated to average .6 hours per response, including the time for reviewing instructions, searching existing data sources, gathering and maintaining the data needed, and completing and reviewing the collection of information. Send comments regarding this burden estimate or any other aspect of this collection of information, including suggestions for reducing the burden, to the Department of Defense, Executive Services and Communications Directorate (0704-0188). Respondents should be aware that notwithstanding any other provision of law, no person shall be subject to any penalty for failing to comply with a collection of information if it does not display a currently valid OMB control number. **PLEASE DO NOT RETURN YOUR FORM TO THE ABOVE ORGANIZATION. SUBMIT COMPLETED FORMS IN ACCORDANCE WITH CONTRACTUAL REQUIREMENTS.**

1. CONTRACTOR	2. CONTRACT	3. PROGRAM	4. REPORT PERIOD
a. NAME	a. NAME	a. NAME	a. FROM *(YYYYMMDD)*
b. LOCATION *(Address and ZIP Code)*	b. NUMBER		b. TO *(YYYYMMDD)*
	c. TYPE / d. SHARE RATIO	b. PHASE *(X one)* RDT&E PRODUCTION	

5. PERFORMANCE DATA

ITEM (1)	CURRENT PERIOD					CUMULATIVE TO DATE					REPROGRAMMING ADJUSTMENTS		AT COMPLETION		
	BUDGETED COST		ACTUAL	VARIANCE		BUDGETED COST		ACTUAL	VARIANCE						
	WORK SCHEDULED (2)	WORK PERFORMED (3)	COST WORK PERFORMED (4)	SCHEDULE (5)	COST (6)	WORK SCHEDULED (7)	WORK PERFORMED (8)	COST WORK PERFORMED (9)	SCHEDULE (10)	COST (11)	COST VARIANCE (12)	BUDGET (13)	BUDGETED (14)	ESTIMATED (15)	VARIANCE (16)
a. ORGANIZATIONAL CATEGORY															
b. COST OF MONEY															
c. GENERAL & ADMINISTRATIVE															
d. UNDISTRIBUTED BUDGET															
e. SUBTOTAL *(Performance Measurement Baseline)*															
f. MANAGEMENT RESERVE															
g. TOTAL															

DD FORM 2734/2, AUG 96   PREVIOUS EDITION MAY BE USED.   LOCAL REPRODUCTION AUTHORIZED.

Reset

Figure 13-8   Cost Performance Report, Format 2—Organizational Category

CLASSIFICATION *(When filled in)*

COST PERFORMANCE REPORT FORMAT 3 - BASELINE DOLLARS IN _____	Form Approved OMB No. 0704-0188

The public reporting burden for this collection of information is estimated to average 6.3 hours per response, including the time for reviewing instructions, searching existing data sources, gathering and maintaining the data needed, and completing and reviewing the collection of information. Send comments regarding this burden estimate or any other aspect of this collection of information, including suggestions for reducing the burden, to the Department of Defense, Executive Services and Communications Directorate (0704-0188). Respondents should be aware that notwithstanding any other provision of law, no person shall be subject to any penalty for failing to comply with a collection of information if it does not display a currently valid OMB control number. **PLEASE DO NOT RETURN YOUR FORM TO THE ABOVE ORGANIZATION. SUBMIT COMPLETED FORMS IN ACCORDANCE WITH CONTRACTUAL REQUIREMENTS.**

1. CONTRACTOR — a. NAME — b. LOCATION *(Address and ZIP Code)*

2. CONTRACT — a. NAME — b. NUMBER — c. TYPE — d. SHARE RATIO

3. PROGRAM — a. NAME — b. PHASE *(X one)* RDT&E / PRODUCTION

4. REPORT PERIOD — a. FROM *(YYYYMMDD)* — b. TO *(YYYYMMDD)*

5. CONTRACT DATA
a. ORIGINAL NEGOTIATED COST | b. NEGOTIATED CONTRACT CHANGES | c. CURRENT NEGOTIATED COST *(a. + b.)* | d. ESTIMATED COST OF AUTHORIZED UNPRICED WORK | e. CONTRACT BUDGET BASE *(c. + d.)* | f. TOTAL ALLOCATED BUDGET | g. DIFFERENCE *(e. - f.)*
h. CONTRACT START DATE *(YYYYMMDD)* | i. CONTRACT DEFINITIZATION DATE *(YYYYMMDD)* | j. PLANNED COMPLETION DATE *(YYYYMMDD)* | k. CONTRACT COMPLETION DATE *(YYYYMMDD)* | l. ESTIMATED COMPLETION DATE *(YYYYMMDD)*

6. PERFORMANCE DATA

ITEM	BCWS CUMULATIVE TO DATE	BCWS FOR REPORT PERIOD	+1	+2	+3	+4	+5	+6	(10)	(11)	(12)	(13)	(14)	UNDISTRIBUTED BUDGET	TOTAL BUDGET
(1)	(2)	(3)	(4)	(5)	(6)	(7)	(8)	(9)	(10)	(11)	(12)	(13)	(14)	(15)	(16)
a. PERFORMANCE MEASUREMENT BASELINE *(Beginning of Period)*															
b. BASELINE CHANGES AUTHORIZED DURING REPORT PERIOD															
c. PERFORMANCE MEASUREMENT BASELINE *(End of Period)*															
7. MANAGEMENT RESERVE															
8. TOTAL															

BUDGETED COST FOR WORK SCHEDULED (BCWS) *(Non-Cumulative)* — SIX MONTH FORECAST — ENTER SPECIFIED PERIODS

DD FORM 2734/3, AUG 96   PREVIOUS EDITION MAY BE USED.   LOCAL REPRODUCTION AUTHORIZED.

Reset

CLASSIFICATION *(When filled in)*

Figure 13-9   Cost Performance Report, Format 3—Baseline

CLASSIFICATION *(When filled in)*

COST PERFORMANCE REPORT
FORMAT 4 - STAFFING

Form Approved
OMB No. 0704-0188

The public reporting burden for this collection of information is estimated to average 5.0 hours per response, including the time for reviewing instructions, searching existing data sources, gathering and maintaining the data needed, and completing and reviewing the collection of information. Send comments regarding this burden estimate or any other aspect of this collection of information, including suggestions for reducing the burden, to the Department of Defense, Executive Services and Communications Directorate (0704-0188). Respondents should be aware that notwithstanding any other provision of law, no person shall be subject to any penalty for failing to comply with a collection of information if it does not display a currently valid OMB control number. **PLEASE DO NOT RETURN YOUR FORM TO THE ABOVE ORGANIZATION. SUBMIT COMPLETED FORMS IN ACCORDANCE WITH CONTRACTUAL REQUIREMENTS.**

1. CONTRACTOR
a. NAME

b. LOCATION *(Address and ZIP Code)*

2. CONTRACT
a. NAME

b. NUMBER

c. TYPE | d. SHARE RATIO

3. PROGRAM
a. NAME

b. PHASE *(X one)*
RDT&E | PRODUCTION

4. REPORT PERIOD
a. FROM *(YYYYMMDD)*

b. TO *(YYYYMMDD)*

5. PERFORMANCE DATA *(All figures in whole numbers)*

ORGANIZATIONAL CATEGORY (1)	ACTUAL CURRENT PERIOD (2)	ACTUAL END OF CURRENT PERIOD (Cumulative) (3)	FORECAST (Non-Cumulative) SIX MONTH FORECAST BY MONTH (Enter names of months) (4)	(5)	(6)	(7)	(8)	(9)	ENTER SPECIFIED PERIODS (10)	(11)	(12)	(13)	(14)	AT COMPLETION (15)
6. TOTAL DIRECT	0	0	0	0	0	0	0	0	0	0	0	0	0	0

DD FORM 2734/4, AUG 96 PREVIOUS EDITION MAY BE USED. | Reset | LOCAL REPRODUCTION AUTHORIZED.

CLASSIFICATION *(When filled in)*

Figure 13-10 Cost Performance Report, Format 4—Staffing

CLASSIFICATION *(When filled in)*

COST PERFORMANCE REPORT FORMAT 5 - EXPLANATIONS AND PROBLEM ANALYSES	*Form Approved* *OMB No. 0704-0188*

The public reporting burden for this collection of information is estimated to average 36.0 hours per response, including the time for reviewing instructions, searching existing data sources, gathering and maintaining the data needed, and completing and reviewing the collection of information. Send comments regarding this burden estimate or any other aspect of this collection of information, including suggestions for reducing the burden, to the Department of Defense, Executive Services and Communications Directorate (0704-0188). Respondents should be aware that notwithstanding any other provision of law, no person shall be subject to any penalty for failing to comply with a collection of information if it does not display a currently valid OMB control number. **PLEASE DO NOT RETURN YOUR FORM TO THE ABOVE ORGANIZATION. SUBMIT COMPLETED FORMS IN ACCORDANCE WITH CONTRACTUAL REQUIREMENTS.**

1. CONTRACTOR	2. CONTRACT	3. PROGRAM	4. REPORT PERIOD
a. NAME	a. NAME	a. NAME	a. FROM *(YYYYMMDD)*
b. LOCATION *(Address and ZIP Code)*	b. NUMBER		b. TO *(YYYYMMDD)*
	c. TYPE / d. SHARE RATIO	b. PHASE *(X one)* ☐ RDT&E ☐ PRODUCTION	

5. EVALUATION

DD FORM 2734/5, AUG 96	PREVIOUS EDITION MAY BE USED.		Reset	Page of Pages
				LOCAL REPRODUCTION AUTHORIZED.

CLASSIFICATION *(When filled in)*

Figure 13-11   Cost Performance Report, Format 5—Explanation and Problem Analysis

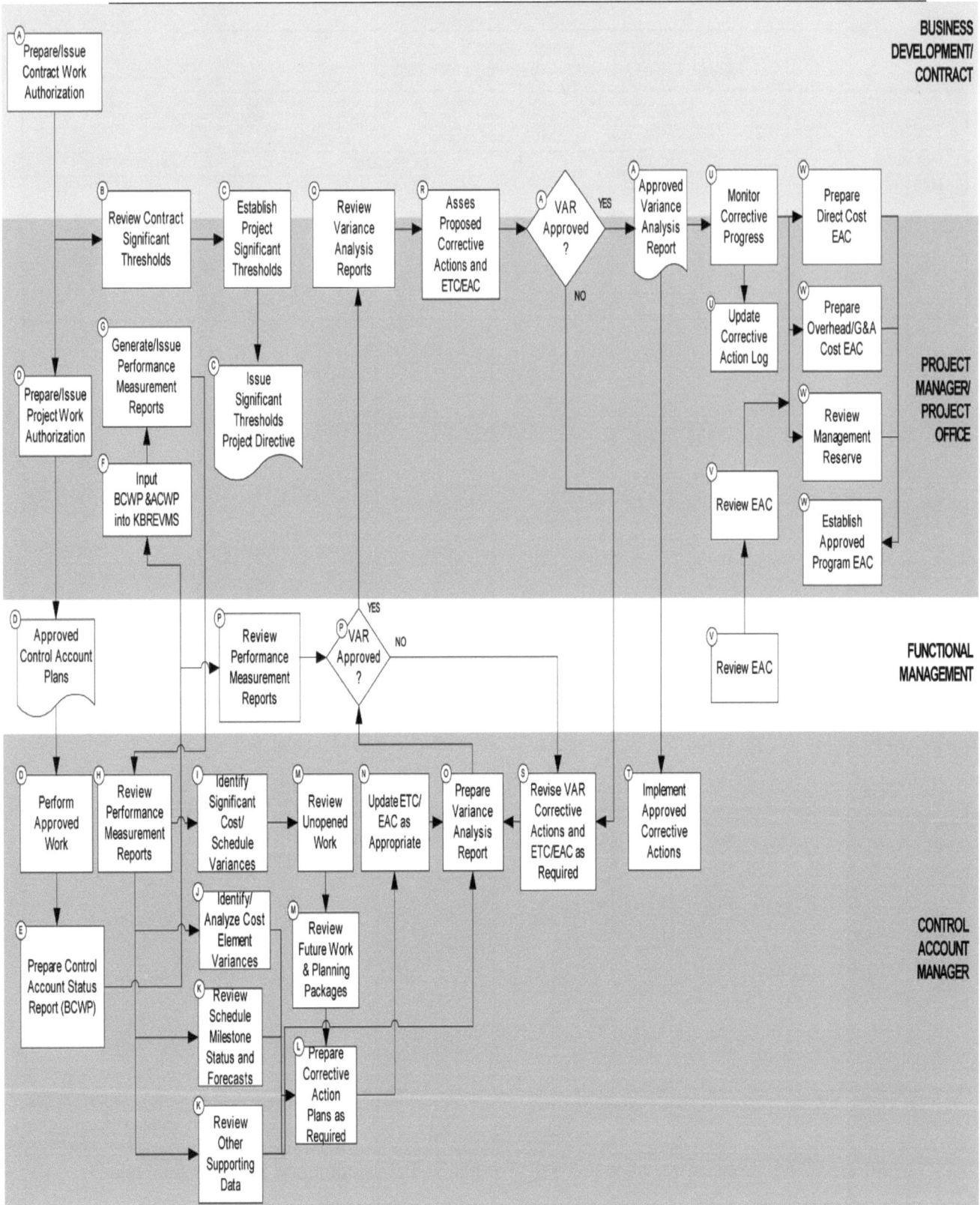

Figure 13-12 Performance Measurement, Analysis, and EAC Update Flowchart

Use or disclosure of data contained on this sheet is
subject to the restriction on the title page of this document.

Page 132

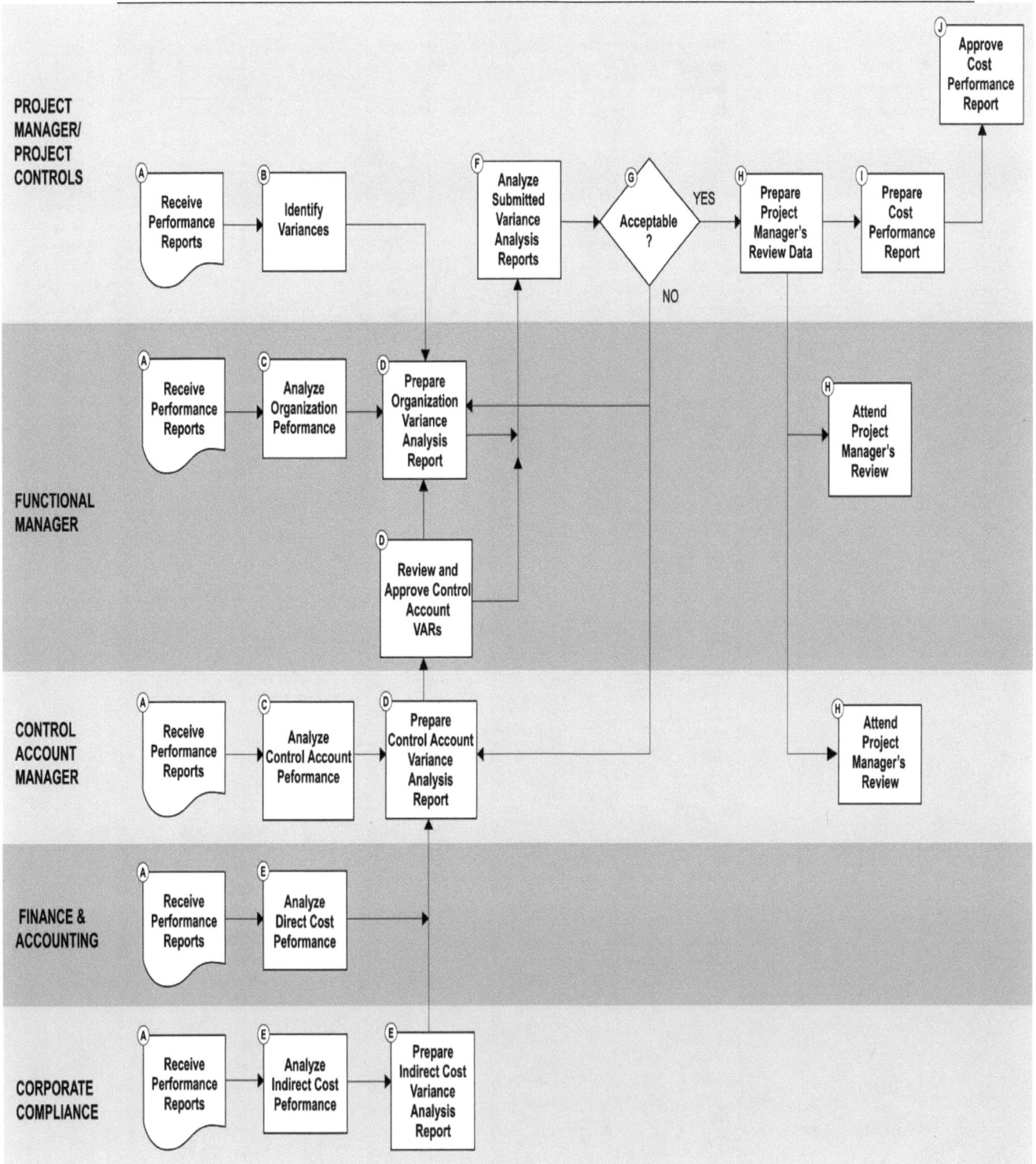

Figure 13-13   Customer Reporting Flowchart

Use or disclosure of data contained on this sheet is
subject to the restriction on the title page of this document.

Page 133

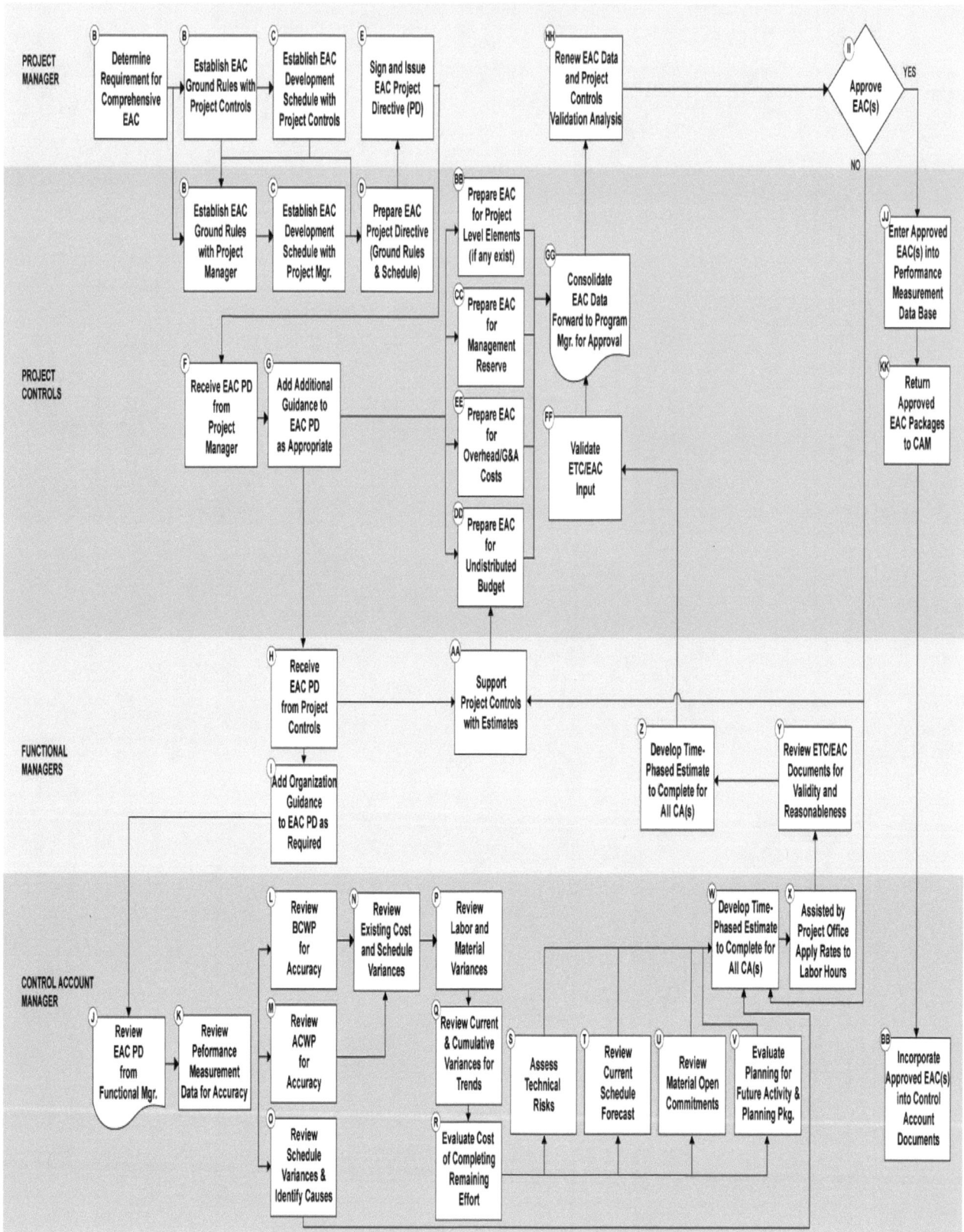

Figure 13-14 Comprehensive EAC Flowchart

Use or disclosure of data contained on this sheet is
subject to the restriction on the title page of this document.

Page 134

14. Revisions and Baseline Control

Use or disclosure of data contained on this sheet is
subject to the restriction on the title page of this document.

Page 135

14.1 Overview

A primary requirement of LGM INTERNATIONAL EVMS is to maintain the validity of the Performance Measurement Baseline (PMB) and CBB throughout the duration of a project. LGM INTERNATIONAL EVMS accomplishes the following:

☐ Incorporates contractual changes in a timely manner, recording the effects of such changes to scope, budgets, schedules, and budget logs.

☐ Reconciles original PMB budgets with current PMB budgets in terms of changes to the authorized work and internal replanning by way of the Change Management Process.

☐ Prohibits retroactive changes to records pertaining to work performed that will change previously reported amounts for direct costs, overhead/G&A, or budgets, except for correction of errors and routine accounting adjustments.

☐ Prevents revisions to the CBB except for customer-directed changes to contractual effort.

• Provides the customer's contracting officer and duly authorized representatives access to all of the LGM INTERNATIONAL EVMS information and supporting documents.

☐ All changes to the Performance Measurement Baseline (PMB) are recorded in Project Change Logs. These changes can include contract changes, reprogramming, internal replanning, use of MR and the application of UB. All PMB changes during the accounting month are reflected in the CPR or C/SSR data submitted to the customer. Narrative Analysis (Format 5 of the CPR) addresses:

  - MR and UB transactions during the month

  - Significant shifts in PMB time phasing and personnel time phasing, as appropriate

  - Notification to the customer of any reprogramming adjustments resulting in an Over Target Baseline (OTB)

Since external and internal cost performance measurement reporting is based on the CBB and the PMB, revisions to the baseline plans must be made according to a formal, documented procedure. This section addresses the process of revisions to the baseline. For internal replanning, a revised CAWA request is the key document for recording such changes for the Change Management Process. In this way, at any stage of a project, the current PMB can be reconciled to the original PMB through the Change Management Process.

14.2 Change Management Process and Documentation

During project execution, events occur which can change either the CBB and the PMB, or just the PMB. LGM INTERNATIONAL EVMS maintains logs which document all such changes to the CBB, PMB and MR. LGM INTERNATIONAL EVMS utilizes a structured Change Management Process to document such changes.

14.2.1 Project Change Logs

LGM INTERNATIONAL maintains logs to document change to the various contract budgets: the CBB log, the UB log, and the MR log. Their purpose is to document transactions affecting the respective budgets.

14.2.1.1 Contract Budget Base Log

Transactions affecting the CBB are recorded in the CBB log. The value for contract changes, whether negotiated or un-priced, is recorded in the CBB log. Changes to the CBB can only be requested by the customer. Figure 14-1 is an example CBB log.

14.2.1.2 Undistributed Budget Log

The UB is the budget for authorized work which has not been distributed at or below the CWBS reporting level to the customer. The UB plus DB is equal to the PMB. The UB exists only for work to be performed in the future. Work that is authorized too late in the current month to be distributed to Control Accounts is reported as UB on the CPR, Format 3. Figure 14-2 is a sample UB log.

Use or disclosure of data contained on this sheet is subject to the restriction on the title page of this document.

Page 136

14.2.1.3 *Management Reserve Log*

MR plus the PMB always equals the CBB. Changes to MR are recorded in the MR log. MR is used to adjust the PMB through internal replanning without affecting the CBB. Figure 14-3 is a sample MR log.

14.3 Contract Changes

Contract changes may be negotiated or un-priced when LGM INTERNATIONAL is directed to incorporate the change. All contract changes will be incorporated into the PMB by the end of the second full reporting period following the negotiation and definitization of the change. Only contractual changes will adjust the CBB.

The contract change process is as follows (see Figure 14-4 for the contract change flowchart):

A. Upon receipt of a contract change (priced or un-priced), Contracts forwards notification of a change to the Project Manager authorizing him or her to expend LGM INTERNATIONAL resources to accomplish the change.

B. The Project Office makes any necessary revisions to contract level documents. This includes the CBB log, the CWBS Index and Dictionary, baseline schedules, and PWAs as described in Paragraphs C through V, below.

C. Immediately upon receipt of the revised CWA, Project Controls enters the value of the change into the CBB log by placing the budget revision amount (excluding fee) shown on the CWA in UB. If the value is negative (i.e., reflecting a negotiation loss for previously un-priced work), UB is reduced and logged accordingly. If there is not enough budget in UB to cover the loss, the PM may reduce MR and update the MR log. The PM may request that CAMs revise their planning to the lower, negotiated value of the work.

D. The Project Office reviews the CWBS, Index and Dictionary and makes any necessary revisions.

E. The Scheduling Section reviews the changes against the PMS and Intermediate Schedules. If changes are necessary, the schedules are revised and formally re-issued.

F. The Project Office identifies the functional organizations affected by the change, and revised schedules are issued to all Functional Managers.

G. The Project Office prepares and issues preliminary PWAs to functional organizations who have no existing PWA. The Project Office prepares preliminary revised PWAs where PWAs exist. For un-priced changes, the PM may choose only to authorize that work which must begin in the near term, i.e., during the next three months.

H. The Functional Managers receive the preliminary or revised PWA and the revised CWBS Index and Dictionary, and schedules. Those Functional Managers who receive a preliminary or revised PWA identify the CAMs affected by the change.

I. The Functional Manager prepares a preliminary CAWA if the new effort is not associated with any existing CAWA. If an appropriate CAWA exists, the Functional Manager prepares a preliminary revised CAWA. These documents are forwarded to the responsible CAM.

J. The CAM receives the preliminary CAWA or preliminary revised CAWA defining the scope of work, budget, and schedule targets.

K. The CAM develops CAPs detailing how the effort will be performed.

L. These plans are negotiated with the Functional Manager.

M. The Functional Manager reviews the CAPs, checking to ensure that the summarized CAPs equal the PWA. If they do, the PWA and accompanying CAPs are forwarded to Project Controls (O). If they do not, the package is forwarded to the Project Manager for negotiation.

N. The Project Manager reviews the PWA and CAPs and either approves or disapproves the change to the PWA. If approved, the PWA and CAPs are forwarded to Project Controls. If disapproved, the PWA and CAPs are returned to the Functional Manager and CAM for revision.

O. Project Controls reviews the PWA and CAPs for consistency, completeness, and conformance to change guidance.

P. Following this revision process, and upon agreement of the scope of work, budget, and schedule, the contract level planning documents are revised if necessary. This includes making any necessary changes to the CWBS, Index and Dictionary, and the PMS and Intermediate Schedules. Note that any revisions made at this time are generally of a clarification nature. These documents always reflect contract requirements. The Project Office develops and issues firm PWAs or revised PWAs.

Q. The CBB, MR and UB logs are annotated to show the formal distribution of budget from contract level UB to the functional organizations.

R   Project Controls enters the CAP information and changes into LGM INTERNATIONAL EVMS baseline database.

S. The functional organization receives the firm PWAs or revised PWAs.

T. The Functional Manager provides final authorization to the CAM via the final CAWA.

U-V. The CAM receives approved CAPs and revised CAWA (U) and proceeds to perform authorized work according to the revised schedule and budget (V).

14.3.1   Negotiated Changes

The overall concept of incorporating a negotiated contract change is identical to the establishment of the original PMB. The CWA is amended, incorporating the contractual changes; as part of the CWA amendment, the CBB is revised, adding or deleting budget per the negotiated change. The executed and released CWA amendment provides authorization from the Project Manager to the Functional Manager and subsequently to the CAMs, to initiate the planning changes required. The PMS is revised as required. Impact on Intermediate/Control Account Schedules, is ascertained, and schedules are revised as necessary. The CWBS Dictionary is revised to incorporate the contract change. Impact on existing Control Accounts and the need for new Control Accounts is discerned by this process.

As the technical, schedule, and budget requirements on new/existing Control Accounts are defined, the Functional Managers/CAMs affected begin planning and budgeting new Control Accounts in detail, and modify existing CAPs as required. Upon approval, the Project Office uses the revised CAPs to update the LGM INTERNATIONAL EVMS database.

The CBB log is utilized to track the budget revisions. The increment of CTC associated with the contract change allocated to MR and the Performance Measurement Baseline is also recorded.

During the interval of time required to revise the CAPs, additions to the CBB will be identified initially as UB. UB will normally be distributed by the end of the second full accounting month following receipt of the contract change. The UB associated with individual contract changes will be tracked until eliminated by issuance to the Control Accounts or MR. This tracing is accomplished by means of the CBB log.

14.3.2   Authorized Un-Priced Changes

The procedures for incorporating authorized un-priced changes are identical to those covered in 14.2.1 above, with the following exceptions:

a. The estimated cost of the undefinitized change will be added to the CBB and Performance Measurement Baseline as UB and reflected on the CBB log.

b. The finalization of CAPs for authorized un-priced changes normally cannot be accomplished in as short a time as for a negotiated contract change. Those Control Accounts which must be added/revised to cover the initial phase of the additional work should be planned and budgeted as soon as possible, and reflected on the CBB log by reducing UB. These Control Accounts will be budgeted at their estimated cost, but no more than 80% of the original UB amount for the change may be distributed until the change has been negotiated.

c. Elimination of UB and firming up the CBB will follow final negotiation of the contract change. All remaining change related UB will normally be distributed by the end of the second full accounting month following negotiation and definitization of the change. If the sum of the current DBs for the authorized/un-priced change exceeds the final negotiated costs of the change, the additional budget will come from a reduction of UB. If there is no UB remaining, the budgets for work not yet started will be reduced.

14.4 Replanning Process

Replanning is a structured process which affects the PMB. In addition to the same stringent controls placed on contract changes, there are other guidelines for replanning.

a. Budget cannot be issued so that the PMB plus MR does not equal the CBB. The CBB includes the cost of all authorized work including LGM INTERNATIONAL estimated cost for authorized/un-priced work.

b. Work Packages cannot be re-planned to cause a slippage in contractual end item deliveries, or to violate Intermediate Schedule guidance.

c. Work Packages must be replanned within the schedule constraints of the CAWA start and finish dates. If these constraints cannot be met, the revised CAWA must state the revised start and/or finish dates of the Control Account.

d. Control Accounts cannot be reopened except for accounting adjustments to ACWP. Such events will adhere to the following guidelines:

   1) The cost details are documented and submitted to the responsible CAM for review.

   2) The CAM will authorize reopening the Control Account for these correcting entries.

   3) The Control Account will be closed again after the entries have been recorded.

e. Transfer of budget from one Control Account to another Control Account must include the associated Scope of Work.

f. MR may be requested by a CAM via a revised CAWA request and must be approved by the Project Manager before MR budget is transferred. This process will be more fully discussed in 14.3.1 below.

g. Any and all revisions to a Control Account (PMB, budget, schedule, or scope of work) initiated by the CAM or Functional Manager must be documented on a revised CAWA request with appropriate approvals. The review of the CAWA helps to ensure that budgets allocated for future efforts are not used for current effort.

The process for documenting and approval of replanning efforts is as follows (see Figure 14-5 for the internally generated change flowchart):

A. Normally, the CAM identifies needed changes to the Control Account. This may be due to changes in resources, availability, estimated quantity changes, usage or efficiency variances. Sometimes, however, the Project Manager, Project Controls or the Functional Manager identify changes and provide guidance to the CAM (by way of meeting, memo, or preliminary revised CAWA) about how to re-plan the CAM's Control Account effort.

B. The CAM revises his or her CAPs and generates a preliminary revised CAWA request. Although the CAM may change the planned schedule dates of tasks, he or she must continue to support Intermediate Schedules and MPSs per current issue. The CAM also requests a log number from Project Controls so that Project Controls can track the status and ultimate disposition of the request.

C. The CAM reviews the requested changes with the Functional Manager.

D. The Functional Manager approves or disapproves the request. If he does not approve the change request, he returns it to the CAM for disposition. The CAM may discard the request, save it for future review or revise and resubmit it. In any case, the CAM will inform Project Controls of its status. If the Functional Manager approves the change request, he assesses the possible Cost and Schedule impacts.

E. The Functional Manager assesses the budget, scope, and schedule impacts. If there are no significant impacts, the revised CAWA is forwarded to Project Controls for updating the progress subsystem in LGM INTERNATIONAL EVMS. It is also returned to the CAM and filed in the EAC backup file. If there are cost and schedule impacts, the Functional Manager prepares a revised PWA for submittal to the Project Manager.

F. Project Controls receives the revised CAWA and updates the progress subsystem in LGM INTERNATIONAL EVMS.

G. Project Controls files the revised CAWA.

H. The Functional Manager prepares a revised PWA request for submittal to the Project Manager. The package will contain the revised PWA, revised CAWA(s) and revised CAP(s).

I. The Project Manager reviews the PWA package with the Functional Manager.

Use or disclosure of data contained on this sheet is subject to the restriction on the title page of this document.

Page 139

J. If the Project Manager denies the request, the PWA package is returned to the Functional Manager. No budget, scope, or schedule changes have been authorized. The Functional Manager returns the revised CAWA(s) and CAP(s) to the CAM(s) for final disposition in the EAC backup file. If he does approve the request, the revised PWA package is forwarded to Project Controls for verification and reissue of a formal PWA and subsequent CAWA(s).

K. Project Controls updates Intermediate Schedules and PMS if necessary.

L. Project Controls reviews the PWA and prepares a revised PWA. Control Accounts are adjusted in the PMB. The revised PWA is returned to the Functional Manager.

M. Project Controls updates the CBB, UB and MR logs, as necessary.

N. The Functional Manager receives the approved, revised PWA with revised schedule guidance.

O. The Functional Manager prepares revised CAWA(s) and forwards the revised and approved CAP(s) to the affected CAM.

P. The CAM receives the revised CAWA and CAP and performs the authorized work according to the revised schedule and budget.

Q. The EAC backup file is the final disposition of all revisions issued to the CAM. The CAM uses the requested and revised CAPs when he is developing an updated EAC.

14.4.1   Internal Replanning

On any project, unexpected events occur regularly during the project cycle. Equipment deliveries, changes in quantity estimates, work method difficulties, subcontractor performance, even internal organization and personnel changes, are a few of the factors that can cause a baseline plan revision. The general rules for replanning are identical to those covered in 14.2 above, except that budget additions to the PMB can only be made from MR, thus not affecting the CBB. The Project Manager may approve the following types of replanning without prior approval of the customer. The customer is notified of significant internal changes by means of the CPR, Baseline Format 3.

Internal Replanning is recorded and documented using revised CAWA requests. This process follows that described in 14.3.

14.4.2   Control Account Scope Change

This type of change may involve transfer of scope, budget, and schedule from one Control Account to another due to reorganization, transfer of personnel, or change in the technical approach. Another reason for this type of change occurs when it is determined that the actual work to be accomplished in a Control Account is beyond the CAWA scope. These types of changes must be PM authorized and within the scope of the authorized contract. If this type of change affects an in-process task, the Work Package PMB budget may be adjusted for future work not in the frozen period (current accounting month) for the additional tasks. If this is clearly impractical, the existing Work Package is closed and a new Work Package is established for the remaining effort. Normal practice in this case is to set BCWS equal to BCWP, transferring the remaining budget to the new Work Package.

14.4.3   Changes to Budget or Schedule to Reflect Efficiency

This type of change is used when it is necessary to re-plan effort within a Control Account in order to reflect a more realistic plan for accomplishing the work. Such a change may involve an early start or a later finish, for instance, or new budget time-phasing. Only unopened WPs may be re-planned without Project Manager approval. Changes may not be made to Work Packages scheduled to begin within the current month without proper justification and documented on a revised CAWA request with Project Manager's approval.

14.4.4   Conversion of Planning Packages into Work Packages

A Planning Package must be converted into Work Package(s) before the associated work can begin. The conversion requires a revised CAWA request but does not require an entry to the CBB log.

14.4.5   Changes Because of Make Versus Buy Decisions

Changes in make versus buy decisions will affect Control Account planning because the budget is changed from one element of cost to another (e.g., labor to material, or vice versa). The Project Office and the CAM must agree to an appropriate adjustment between the amount of budget returned from one element of cost and the amount gained in the other element of cost. Due to the different overhead and ODC adjustments, the amount of the labor budget given up may not equal the material budget gained. The Project Office and the CAM should also consider the estimated cost of the change approach. There is normally little or no effect to the total PMB for this type of change. This type of change must be documented on a revised CAWA request and may also be recorded on the UB or MR logs, if required.

14.4.6   Replanning Work Packages

Once Work Packages have been defined and budgeted, internal controls are established to minimize further changes to budgets, schedule, or scope of work. The following controls are used:

a. Future unopened Work Packages and the remaining open portion of a Work Package beyond the current accounting month may be re-planned.

b. Work Package cost and schedule baselines in the current month and past cannot be re-planned.

c. Replanning of future Work Packages is acceptable when:

   - The start date is beyond the frozen period (current accounting month) and,

   - The schedule start and completion dates are within the constraints of higher level schedule controlling milestones and,

   - The Control Account scheduled completion date is not impacted and,

   - It is documented on a revised CAWA and approved.

14.5   Replanning Requiring Project Manager Approval

In addition to those changes discussed previously, changes of a larger scale may be made with Project Manager approval and customer notification. These changes normally affect a large percentage of existing Control Accounts. Customer notification is required prior to the following type of changes being implemented

14.5.1   Major Replanning

When budgets and/or schedules for existing Control Accounts are determined to no longer be a realistic baseline against which to measure performance, the Project Manager will notify the customer prior to implementing a major replanning. This may involve making changes to many in-process Work Packages, but does not involve an Over Target Baseline (OTB). LGM INTERNATIONAL would still plan to come within the target cost in this situation.

14.5.2   Formal Reprogramming

Reprogramming means establishing a PMB in excess of the CBB. This situation can occur when a contract overrun is projected and LGM INTERNATIONAL has determined that the current PMB is an unrealistic measurement of remaining work effort. In such a case, the customer is formally notified of the OTB and internal plans are revised as required to reflect a more realistic plan for completing the project. The techniques and documentation are similar to those discussed previously, with the exception that schedule variances currently existing in open Control Accounts may be eliminated after notification of cognizant customer personnel. Open Work Packages may be closed (if no longer applicable) or revised, and BCWS/BCWP adjusted so that for the Control Account as a whole, BCWS=BCWP at the project cutoff date. All changes will be accomplished via revised CAWA so that a clear, auditable trail of all adjustments is maintained. The CBB log will reflect the addition to the CBB, and all budget changes to affected Control Accounts are recorded on revised CAWAs.

Use or disclosure of data contained on this sheet is
subject to the restriction on the title page of this document.

Page 141

14.6   Retroactive Adjustments

Should such corrections result in a significant distortion of previously reported performance, a thorough explanation will be provided to internal management and to the customer via CPR Format 5. Retroactive changes to BCWS and BCWP are prohibited. The same restriction refers to actual costs with exceptions as noted previously. Such corrections are made as current period adjustments.

Use or disclosure of data contained on this sheet is subject to the restriction on the title page of this document.

Page 142

Report : CBB_LOG   **Contract Budget Baseline Log**   **Date :**   **05/17/2004**

File :   **DEMOADV**   **XYZ Building**

Cost Account	Date	Ref #	Chg #	Description	Contract Target Cost	Authorized Unpriced Work	Contract Budget Base	Management Reserve	Performance Measurement Baseline	Undistributed Budget	Distributed Budget
	04/12/1998	238		Set Baseline	2210000.00	0.00	2210000.00	30000.00	2180000.00	2180000.00	0.00
1.1.1.1/1400	04/12/1998	238		Set Baseline	0.00	0.00	0.00	0.00	0.00	-75862.06	75862.06
1.1.1.2/1420	04/12/1998	238		Set Baseline	0.00	0.00	0.00	0.00	0.00	-34288.91	34288.91
1.1.2.1/1600	04/12/1998	238		Set Baseline	0.00	0.00	0.00	0.00	0.00	-32693.78	32693.78
1.1.2.2/1600	04/12/1998	238		Set Baseline	0.00	0.00	0.00	0.00	0.00	-3802.36	3802.36
1.2.1.1/1332	04/12/1998	238		Set Baseline	0.00	0.00	0.00	0.00	0.00	-153249.27	153249.27
1.2.1.2/1331	04/12/1998	238		Set Baseline	0.00	0.00	0.00	0.00	0.00	-73728.92	73728.92
1.2.3/1000	04/12/1998	238		Set Baseline	0.00	0.00	0.00	0.00	0.00	-223429.85	223429.85
1.3.1/1220	04/12/1998	238		Set Baseline	0.00	0.00	0.00	0.00	0.00	-191730.62	191730.62
1.3.2/1220	04/12/1998	238		Set Baseline	0.00	0.00	0.00	0.00	0.00	-35302.40	35302.40
1.4/1500	04/12/1998	238		Set Baseline	0.00	0.00	0.00	0.00	0.00	-45339.90	45339.90
1.6/1000	04/12/1998	238		Set Baseline	0.00	0.00	0.00	0.00	0.00	-1307734.85	1307734.85
	04/12/1998			Current Balances:	2210000.00	0.00	2210000.00	30000.00	2180000.00	2837.08	2177162.92
	05/15/1998	241		New work scope	80000.00	0.00	80000.00	0.00	80000.00	80000.00	0.00
	05/15/1998	242		Allocate MR for new scope	0.00	0.00	0.00	-10000.00	10000.00	10000.00	0.00
1.1.1.2/1420	05/15/1998	242		New work scope	0.00	0.00	0.00	0.00	0.00	-10907.34	10907.34
1.1.2.1/1600	05/15/1998	242		New work scope	0.00	0.00	0.00	0.00	0.00	-25161.41	25161.41
1.1.2.2/1600	05/15/1998	242		New work scope	0.00	0.00	0.00	0.00	0.00	-539.00	539.00
1.2.1.1/1332	05/15/1998	242		New work scope	0.00	0.00	0.00	0.00	0.00	-34544.63	34544.63
1.3.1/1220	05/15/1998	242		New work scope	0.00	0.00	0.00	0.00	0.00	-20128.99	20128.99
	05/15/1998			Current Balances:	2290000.00	0.00	2290000.00	20000.00	2270000.00	1555.71	2268444.29
1.1.2.2/1600	06/06/1998	244		Added Ergonomic engineer for lab	0.00	0.00	0.00	0.00	0.00	-1420.33	1420.33
	06/06/1998			Current Balances:	2290000.00	0.00	2290000.00	20000.00	2270000.00	135.38	2269864.62

Data agrees with program header information

Cobra (R) by WST Corp.   Page :   1

Figure 14-1   Sample Contract Budget Base Log

Report : UB_LOG **Undistributed Budget Log** **Date :05/17/2004**

File : DEMOADV **XYZ Building**

Cost Account	Date	Ref#	Chg#	Description	Increase	Decrease	Balance
	04/12/1998	238		Set Baseline	2180000.00		2180000.00
1.1.1.1/1400	04/12/1998	238		Set Baseline		75862.06	2104137.94
1.1.1.2/1420	04/12/1998	238		Set Baseline		34288.91	2069849.03
1.1.2.1/1600	04/12/1998	238		Set Baseline		32693.78	2037155.25
1.1.2.2/1600	04/12/1998	238		Set Baseline		3802.36	2033352.89
1.2.1.1/1332	04/12/1998	238		Set Baseline		153249.27	1880103.62
1.2.1.2/1331	04/12/1998	238		Set Baseline		73728.92	1806374.70
1.2.3/1000	04/12/1998	238		Set Baseline		223429.85	1582944.85
1.3.1/1220	04/12/1998	238		Set Baseline		191730.62	1391214.23
1.3.2/1220	04/12/1998	238		Set Baseline		35302.40	1355911.83
1.4/1500	04/12/1998	238		Set Baseline		45339.90	1310571.93
1.6/1000	04/12/1998	238		Set Baseline		1307734.85	2837.08
	05/15/1998	241		New work scope	80000.00		82837.08
	05/15/1998	242		Allocate MR for new scope	10000.00		92837.08
1.1.1.2/1420	05/15/1998	242		New work scope		10907.34	81929.74
1.1.2.1/1600	05/15/1998	242		New work scope		25161.41	56768.33
1.1.2.2/1600	05/15/1998	242		New work scope		539.00	56229.33
1.2.1.1/1332	05/15/1998	242		New work scope		34544.63	21684.70
1.3.1/1220	05/15/1998	242		New work scope		20128.99	1555.71
1.1.2.2/1600	06/06/1998	244		Added Ergonomic engineer for lab		1420.33	135.38

Cobra (R) by WST Corp. **Page : 1**

Figure 14-2 Sample Undistributed Budget Log

Cost Account	Date	Ref #	Chg #	Description	Increase	Decrease	Balance
	04/12/1998	238		Set Baseline	30000.00		30000.00
	05/15/1998	242		Allocate MR for new scope		10000.00	20000.00

Figure 14-3   Sample Management Reserve Log

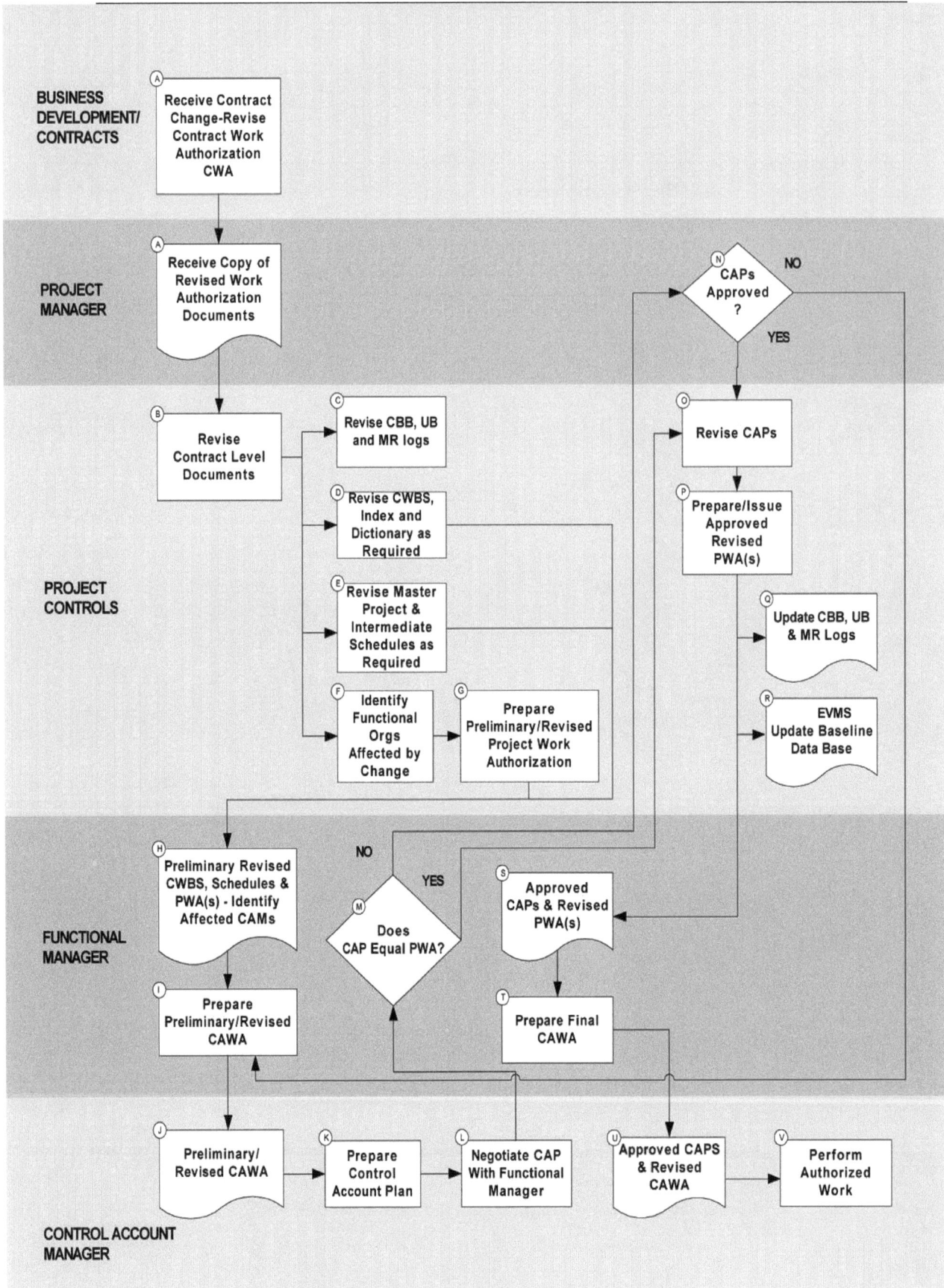

Figure 14-4   Contract Change Flowchart

Use or disclosure of data contained on this sheet is
subject to the restriction on the title page of this document.

Page 146

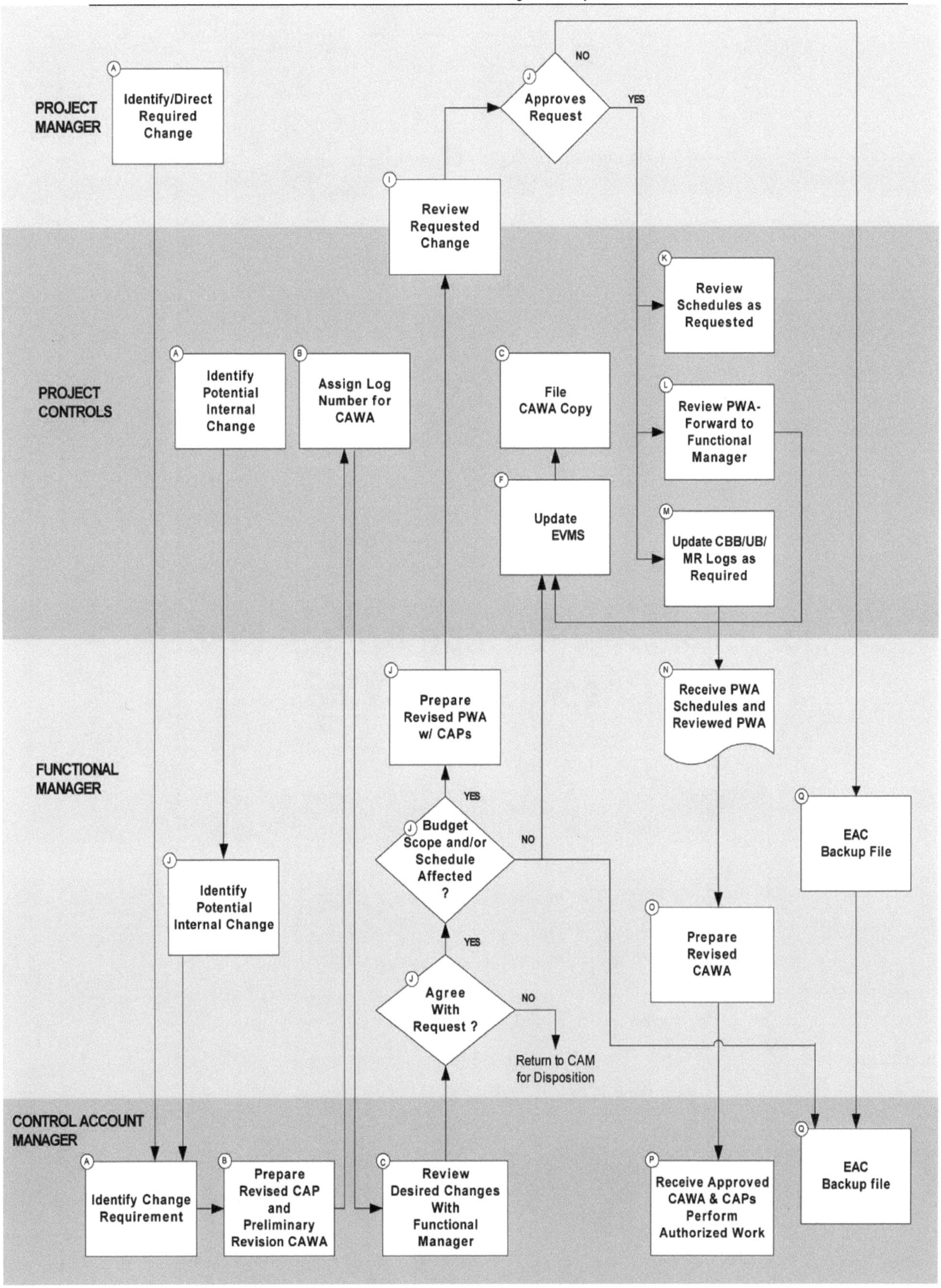

Figure 14-5   Internally Generated Change Flowchart

Use or disclosure of data contained on this sheet is
subject to the restriction on the title page of this document.

Page 147

15. Evaluation/Demonstration Review Checklist for EVMS

se or disclosure of data contained on this sheet is
subject to the restriction on the title page of this document.

Page 148

15.1 Criteria Checklist Cross Reference

The EVMS Criteria Checklist is annotated with the applicable LGM INTERNATIONAL EVMS system description paragraph and figure numbers.

The checklist is to be reviewed at inception and during the continual surveillance program of the EVMS program.

LGM INTERNATIONAL EARNED VALUE MANAGEMENT SYSTEM IMPLEMENTATION SELF-ASSESSMENT

The following is a list of the 32 criteria that defined and required to be met under a formal certified Earned Value Management System (EVMS).

Note: There are five main categories which each of the 32 criteria fall under: Organization; Planning, Scheduling, and Budgeting; Accounting Considerations; Analysis and Management; Revisions and Data Maintenance

1. ORGANIZATION

1.1 Define the authorized work elements for the program. A work breakdown structure (WBS), tailored for effective internal management control is commonly used in the process.
RESULT: MEETS STANDARD ☐ NOT DEMONSTRATED ☐   ACTION ITEM ☐
Requirement / Check: WBS structure and dictionary
LGM INTERNATIONALI SD: 4.2.1; 4.2.1.1; 4.2.1.2; 4.2.2; 4.2.3; 4.2.4

1.2 Identify the program organizational structure including the major subcontractors   responsible for accomplishing the authorized work, and define the organizational elements in which work will be planned and controlled.
RESULT: MEETS STANDARD ☐   NOT DEMONSTRATED ☐   ACTION ITEM ☐
Requirement / Check: OBS structure and/or Organizational Chart
LGM INTERNATIONALI SD: 4.3.1; 4.3.2; 4.3.3; 4.3.3.1; 4.3.3.2; 4.3.3.5; 4.3.3.6

1.3 Provide for the integration of the company's planning, scheduling, budgeting, work authorization and cost accumulation processes with each other, and as appropriate, the program work breakdown structure and the program organizational structure.
RESULT: MEETS STANDARD ☐   NOT DEMONSTRATED ☐   ACTION ITEM ☐
Requirement / Check: Contract and/or Modification(s), RAM, CAP, CAWA, Cost loads
LGM INTERNATIONALI SD: 4.6; 4.7.1

1.4 Identify the company organization or function responsible for controlling overhead (indirect costs).
RESULT: MEETS STANDARD ☐ NOT DEMONSTRATED ☐   ACTION ITEM ☐
Requirement / Check: Government / Corporate Compliance showing current forecasted OH rates, DCMA letter showing billing rates.
LGM INTERNATIONALI SD: 5.5.H.1; 11.5.4; 12.1; 12.2.1; 12.2.2; 12.4

1.5 Provide for integration of the program work breakdown structure and the program organizational structure in a manner that permits cost and schedule performance measurement by elements of either or both structures as needed.
RESULT: MEETS STANDARD ☐ NOT DEMONSTRATED ☐   ACTION ITEM ☐
Requirement / Check: RAM
LGM INTERNATIONALI SD: 4.4.1; 4.5

2. PLANNING, SCHEDULING, & BUDGETING

2.1 Schedule the authorized work in a manner which describes the sequence of work and identifies significant task interdependencies required to meet the requirements of the program.
RESULT: MEETS STANDARD ☐   NOT DEMONSTRATED ☐   ACTION ITEM ☐
Requirement / Check: Schedule, cost loads (Primavera to EVMS), CAWA, CAP
LGM INTERNATIONALI SD: 6.3; 6.4.2; 6.6

2.2 Identify physical products, milestones, technical performance goals, or other indicators that will be used to measure progress.
RESULT: MEETS STANDARD ☐   NOT DEMONSTRATED ☐   ACTION ITEM ☐
                Requirement / Check: CAP, cost loads (Primavera to EVMS), Primavera, Schedule
                LGM INTERNATIONALI SD: 6.4.1; 6.6

2.3 Establish and maintain a time-phased budget baseline, at the control account level, against which program performance can be measured. Initial budgets established for performance measurement will be based on either internal management goals or the external customer negotiated target cost including estimates for authorized but undefinitized work. Budget for far-term efforts may be held in higher level accounts until an appropriate time for allocation at the control account level. On government contracts, if an over-target baseline is used for performance measurement reporting purposes prior notification must be provided to the customer.
RESULT: MEETS STANDARD ☐   NOT DEMONSTRATED ☐   ACTION ITEM ☐
                Requirement / Check: CAP (detail and summary)
                LGM INTERNATIONALI SD: 5.3; 5.4; 5.6; 5.10; 5.11

2.4 Establish budget for authorized work with identification of significant cost elements (labor, material, subcontracts, other directs, etc.) as needed for internal management and for control of subcontractors.
RESULT: MEETS STANDARD ☐   NOT DEMONSTRATED ☐   ACTION ITEM ☐
                Requirement / Check: CAP, CAWA
                LGM INTERNATIONALI SD: 4.7.4.1; 5.3; 7.1; 7.2; 7.2.2; 7.2.3; 8.3

2.5 To the extent it is practical to identify the authorized work in discrete work packages, establish budgets for this work in terms of dollars, hours, or other measurable units and where the entire control account is not subdivided into work packages, identify the far-term effort in larger planning packages for budget and scheduling purposes.
RESULT: MEETS STANDARD ☐   NOT DEMONSTRATED ☐   ACTION ITEM ☐
                Requirement / Check: CAP
                LGM INTERNATIONALI SD: 7.2.3; 8.3

2.6 Provide that the sum of all work package budgets plus planning package budgets within a control account equals the control account budget.
RESULT: MEETS STANDARD ☐ NOT DEMONSTRATED ☐   ACTION ITEM ☐
      Requirement / Check: CAP
      LGM INTERNATIONALI SD: 5.3; 5.5

2.7 Identify and control level of effort (LOE) activity by time-phased budgets established for this purpose. Only that effort which is immeasurable or for which measurement is impractical may be classified as LOE.
RESULT: MEETS STANDARD ☐   NOT DEMONSTRATED ☐   ACTION ITEM ☐
                Requirement / Check: CAP
                LGM INTERNATIONALI SD: 4.7.1; 4.7.3.2; 4.7.4.3; 8.4.3

2.8 Establish overhead budgets for each significant organizational component of the company for expenses which will become indirect cost. Reflect in the program budgets, at the appropriate level, the amounts in overhead pools that are planned to be allocated to the program as indirect costs.
RESULT: MEETS STANDARD ☐   NOT DEMONSTRATED ☐   ACTION ITEM ☐

  Requirement / Check: Government Compliance letter showing current forecasted OH rates, DCMA letter showing billing rates
                LGM INTERNATIONALI SD: 11.5; 12.2.1; 12.2.2; 12.2.3; 12.4

2.9 Identify management reserves and undistributed budget.
RESULT: MEETS STANDARD ☐ NOT DEMONSTRATED ☐   ACTION ITEM ☐
                Requirement / Check: Contract Budget Baseline (CBB) log, CPR format #1
                LGM INTERNATIONALI SD: 5.3; 5.4; 5.5; 5.8

2.10 Provide that the program target cost is reconciled with the sum of all internal program budgets and management reserves.
RESULT: MEETS STANDARD☐   NOT DEMONSTRATED ☐ACTION ITEM ☐
Requirement / Check: CPR format #1 & by control account, CBB log
LGM INTERNATIONALI SD: 5.4; 5.5

3. ACCOUNTING CONSIDERATIONS

3.1 Record direct cost in a manner consistent with the budgets in a formal system controlled by the general books of account.
RESULT: MEETS STANDARD☐   NOT DEMONSTRATED☐   ACTION ITEM☐
Requirement / Check: CAP
LGM INTERNATIONALI SD: 11.5; 11.5.1; 11.5.2; 11.5.3; 11.5.4

3.2 When a work breakdown structure is used, summarize direct costs from control accounts into the work breakdown structure without allocation of a single control account to two or more work breakdown structure elements.
RESULT: MEETS STANDARD ☐ NOT DEMONSTRATED ☐ ACTION ITEM☐
Requirement / Check: WBS structure – looking for pure integration, no low-level cost should roll-up to multiple higher-level accounts and no cost duplication should be present.
LGM INTERNATIONALI SD: 11.3

3.3 Summarize direct costs from the control accounts into the contractor's organizational elements without allocation of a single control account to two or more organizational elements.
RESULT: MEETS STANDARD ☐NOT DEMONSTRATED ☐ ACTION ITEM☐
Requirement / Check: RAM, Organizational Chart
LGM INTERNATIONALI SD: 11.3

3.4 Record all indirect costs which will be allocated to the contract.
RESULT: MEETS STANDARD☐   NOT DEMONSTRATED ☐ACTION ITEM☐
Requirement / Check: WBS structure and dictionary
LGM INTERNATIONALI SD: 11.5.4; 12.2.3; 12.2.3.4

3.5 Identify unit costs, equivalent unit costs, or lot costs when needed.
RESULT: MEETS STANDARD☐   NOT DEMONSTRATED ☐ ACTION ITEM☐
Requirement / Check: Estimate, PCMS
LGM INTERNATIONALI SD: 11.4.3

3.6 For Earn Value Management System (EVMS), the material accounting system will provide for:
a. Accurate cost accumulation and assignment of costs to control accounts in a manner consistent with the budgets using recognized, acceptable, costing techniques.
RESULT: MEETS STANDARD☐   NOT DEMONSTRATED☐ ACTION ITEM ☐
Requirement / Check: Procurement Report, Material Status Report (MSR), P3, PCMS,
MDCS LGM INTERNATIONALI SD: 9.2; 9.3.1; 9.4.3.4; 9.4.3.5; 9.4.3.6; 9.6

b. Cost performance measurement at the point in time most suitable for the category of material involved, but no earlier than the time of progress payments or actual receipt of material.
RESULT: MEETS STANDARD ☐NOT DEMONSTRATED☐   ACTION ITEM☐
Requirement / Check: Material Status Report (MSR), P6, Procurement Management Control System, Material Data Control System

LGM INTERNATIONALI SD: 9.2; 9.3.1; 9.4.3.4; 9.4.3.5; 9.4.3.6; 9.6

c. Full accountability of all material purchased for the program including the residual inventory.
RESULT: MEETS STANDARD ☐NOT DEMONSTRATED☐   ACTION ITEM☐
Requirement / Check: Material Status Report (MSR), P6, Procurement Management Control System, Material Data Control System

LGM INTERNATIONALI SD: 9.5

4. ANALYSIS & MANAGEMENT REPORTS

4.1 At least on a monthly basis, generated the following information at the control account and other levels as necessary for management control using actual cost data from, or reconcilable with, the accounting system.
a. Comparison of the amount of planned budget and the amount of budget earned for work accomplished. This comparison provides the schedule variance.
RESULT: MEETS STANDARD ☐ NOT DEMONSTRATED ☐ ACTION ITEM ☐
        Requirement / Check: CPR format #1, VAR
        LGM INTERNATIONALI SD: 13.1; 13.2; 13.4.2.2

b. Comparison of the amount of the budget earned and the actual (applied where appropriated) direct costs for the same work. This comparison provides the cost variance.
RESULT: MEETS STANDARD ☐ NOT DEMONSTRATED ☐ ACTION ITEM ☐
        Requirement / Check: CPR format #1, VAR
        LGM INTERNATIONALI SD: 13.1; 13.2; 13.4.2.2

4.2 Identify, at least monthly, the significant differences between both planned and actual schedule performance and planned and actual cost performance and provide the reasons for the variance in the detail needed by program management.
RESULT: MEETS STANDARD ☐ NOT DEMONSTRATED ☐ ACTION ITEM ☐
        Requirement / Check: VAR
        LGM INTERNATIONALI SD: 13.4.2.1; 13.4.2.2

4.3 Identify budgeted and applied (or actual) indirect costs at the level and frequency needed by management for effective control, along with the reasons for any significant variances.
RESULT: MEETS STANDARD ☐ NOT DEMONSTRATED ☐ ACTION ITEM ☐
  Requirement / Check: Government / Corporate Compliance showing current forecasted OH rates, DCMA letter showing billing rates.
        LGM INTERNATIONALI SD: 12.4

4.4 Summarize the data elements and associated variances through the program organization and/or work breakdown structure to support management needs and any customer reporting specified in the contract.
RESULT: MEETS STANDARD ☐ NOT DEMONSTRATED ☐ ACTION ITEM ☐
        Requirement / Check: PCMS, CAP, VAR
        LGM INTERNATIONALI SD: 13.1; 13.3.2; 13.6.1; 13.6.2; 13.6.2.1

4.5 Implement managerial actions taken as the result of earned value information.
RESULT: MEETS STANDARD ☐ NOT DEMONSTRATED ☐ ACTION ITEM ☐
        Requirement / Check: CPR format #1, VAR Corrective Action log
        LGM INTERNATIONALI SD: 13.4.3; 13.5

4.6 Develop revised estimates of cost at completion based on performance to date, commitment values for material, and estimates of future conditions. Compare this information with the performance measurement baseline to identify variances at completion important to company management and any applicable customer reporting requirements including statements of funding requirements.
RESULT: MEETS STANDARD ☐ NOT DEMONSTRATED ☐ ACTION ITEM ☐
        Requirement / Check: CPR format #1 and/or Contract Funds Status Report
LGM INTERNATIONALI SD: 13.5; 13.5.1; 13.5.2; 13.6.2; 13.6.2.1; 13.6.2.2; 13.6.2.3; 13.7.1; 13.7.3

5. REVISIONS & DATA MAINTENANCE

5.1 Incorporate authorized changes in a timely manner, recording the effects of such changes in budgets and schedules. In the directed effort prior to negotiation of a change, base such revision on the amount estimated and budgeted to the program organizations.
RESULT: MEETS STANDARD ☐ NOT DEMONSTRATED ☐ ACTION ITEM ☐
        Requirement / Check: CBB log, Contract and/or Modification(s)
        LGM INTERNATIONALI SD: 14.1; 14.3; 14.3.1; 14.3.2

5.2 Reconcile current budgets to prior budgets in terms of changes to the authorized work and internal replanning in the detail needed by management for effective control.
RESULT: MEETS STANDARD ☐ NOT DEMONSTRATED ☐ ACTION ITEM ☐
                    Requirement / Check: CBB log, Contract and/or Modification(s)
                    LGM INTERNATIONALI SD: 14.1; 14.2.1.1; 14.3; 14.3.1

5.3 Control retroactive changes to records pertaining to work performed that would change previously reported amounts for actual cost, earned value, or budgets. Adjustments should be made only for correction of errors, routine accounting adjustments, effects of customer or management directed changes, or to improve the baseline integrity and accuracy of performance measurement data.
RESULT: MEETS STANDARD ☐ NOT DEMONSTRATED ☐ ACTION ITEM ☐
                    Requirement / Check: CBB log
                    LGM INTERNATIONALI SD: 14.1; 14.4

5.4 Prevent revisions to the program budget except for authorized changes.
RESULT: MEETS STANDARD ☐ NOT DEMONSTRATED ☐ ACTION ITEM ☐
                    Requirement / Check: CBB log, PWA, CPR format #1
                    LGM INTERNATIONALI SD: 14.1; 14.2.1.1; 14.3; 14.3.1; 14.4; 14.4.1

5.5 Document changes to the performance measurement baseline.
RESULT: MEETS STANDARD ☐ NOT DEMONSTRATED ☐ ACTION ITEM ☐
                    Requirement / Check: CBB log, PWA
                    LGM INTERNATIONALI SD: 14.2.1; 14.2.1.1; 14.2.1.2; 14.2.1.3; 14.3

www.ingramcontent.com/pod-product-compliance
Lightning Source LLC
Chambersburg PA
CBHW041710210326
41598CB00007B/597